MINNESOTA UNDERGROUND

A GUIDE TO CAVES & KARST, MINES & TUNNELS

SECOND EDITION

Other books by Doris Green

Elsie's Story: Chasing a Family Mystery

Explore Wisconsin Rivers

Minnesota Underground and the Best of the Black Hills (1st edition)

Wisconsin Underground: A Guide to Caves, Tunnels, and Mines
Around the Badger State

Other books by Greg Brick:

Iowa Underground: A Guide to the State's Subterranean Treasures

Subterranean Twin Cities

Minnesota Caves: History and Lore

MINNESOTA UNDERGROUND

A GUIDE TO CAVES & KARST, MINES & TUNNELS

SECOND EDITION

DORIS GREEN & GREG BRICK, PhD

HENSCHELHAUS PUBLISHING, INC.
MILWAUKEE, WISCONSIN

HenschelHAUS Publishing, Inc.
2625 S. Greeley St. Suite 201
Milwaukee, WI 53207
www.henschelHAUSbooks.com

HenschelHAUS books may be purchased for educational, business, or sales promotional use. For information, please email info@henschelHAUSbooks.com

ISBN: 978159598-746-4
E-ISBN: 978159598-747-1
LCCN: 2019947835

Publisher's Cataloging-In-Publication Data
(Prepared by The Donohue Group, Inc.)

Names: Green, Doris (Doris M.), author. | Brick, Greg A., author.
Title: Minnesota underground : a guide to caves & karst, mines & tunnels / Doris Green & Greg Brick, PhD.
Description: Second edition. | Milwaukee, Wisconsin : HenschelHAUS Publishing, Inc., [2019] | Includes bibliographical references.
Identifiers: ISBN 9781595987464 | ISBN 9781595987471 (ebook)
Subjects: LCSH: Caves--Minnesota--Guidebooks. | Karst--Minnesota--Guidebooks. | Mines and mineral resources--Minnesota--Guidebooks. | Minnesota--Description and travel.
Classification: LCC GB605.M6 G744 2019 (print) | LCC GB605.M6 (ebook) | DDC 551.44/7/09776--dc23

Cover design by Lisa Imhoff, Grey Horse Studio
Front cover photo: Mystery Cave, reprinted with permission from the MN Department of Natural Resources.

"Many subterranean palaces are said to be found in Minnesota."
—Fredrika Bremer, *The Homes of the New World*, 1853

TABLE OF CONTENTS

SOUTHWESTERN MINNESOTA

MINNEAPOLIS-ST. PAUL METRO AREA AND ST. CLOUD

ACKNOWLEDGMENTS

First and foremost, thank you to Kira Henschel of HenschelHaus Publishing for suggesting this second edition of *Minnesota Underground*.

Thanks also to John Ackerman, founder of the Minnesota Cave Preserve; photographer Tony Andrea, East End Productions; Dave Battistel, Gunflint Range railroad and Paulson Mine historian; Mark Bishop, owner of Niagara Cave; Jeff Boland, outreach and history education coordinator at Eagle Bluff Environmental Learning Center; Sandy Brandley, board member with Ely Greenstone Public Art, the nonprofit manager of the Ely Arts & Heritage Center in the historic Pioneer Mine buildings; Karen Cooper, urbancreek.com; Shawnee Hoffman, photographer; Keri Huber, Minnesota State Fair archivist; Jim Keenan, past owner of Ye Old Mill; Janet Wildung Lanphere, executive director, Luverne Area Chamber of Commerce; Hannah Lieffring, Minnesota Speleological Survey newsletter editor; Jessica Madole and Zach Nugent, Minnesota Zoo; Eric McMaster, owner of Wisconsin's Crystal Cave; Beth Pierce, director, Iron Range Tourism Bureau; James Pointer, interpretive supervisor at Lake Vermilion-Soudan Underground Mine State Park; Sue Prom, Voyageur Canoe Outfitters at the Gunflint Trail; Debra Richardson, executive director, Fillmore County History Center; Deborah Rose, photographer with the Minnesota Department of Natural Resources; Julie Rubel, owner of Iowa's Crystal Lake Cave; Christine Salomon, associate professor with the University of Minnesota School of Pharmacy; Reva Sehr, administrative assistant with the Luverne Area Chamber of Commerce and others with Hinkly House; Gordon Smith, National Cave Museum, for generous access to his vast cave postcard collection; and Gary Soule, Wisconsin Speleological Society cave archivist and historian.

Doris also thanks Michael H. Knight for gamely accompanying her on treks to southeastern Minnesota and the Iron Range. Greg would like to thank his patient wife Cindy and cat Tori for their support when literary deadlines loom.

INTRODUCTION

When the first edition of *Minnesota Underground* hit bookstores in 2003, white nose syndrome had not yet begun to decimate the hibernating bat populations across the eastern United States, never mind Minnesota. Taconite pellets still rolled, marble-like, beneath the feet of visitors at several Mesabi Range overlooks. While dye tracings revealed the paths of various underground streams, few educational centers had yet to tackle in depth the topic of karst education.

Much has changed in 16 years: Nature has reclaimed several pit mine sites; museums have developed ever more realistic replica cave and mine passages; and awareness of the need for karst education and groundwater protection has grown. The Minnesota Department of Natural Resources has preserved additional public lands and installed bat gates to protect the state's hibernating bats. The Minnesota Speleological Survey and the Minnesota Caving Club, both grottos (member groups) of the National Speleological Society (www.caves.org), have worked with the DNR to explore, preserve, and study the state's subterranean treasures. This edition reflects these changes.

This edition also offers an expanded Minnesota sites list, with descriptions of 82 sites spread through 23 counties. Many natural caves and other karst features are found in Fillmore County, while manmade caves are centered around the Twin Cities. Several mine-related sites follow U.S. Highway 169 through the Iron Range and, a bit to the east, I-35 and State Highway 61 lead to a diversity of sites near the North Shore. Directions, seasons/hours, precautions, and amenities, where applicable, are offered for most sites. Twenty-two sidebars add surprising background and history too good to omit—from scimitar cats to the Gainey Gold Mine fraud, from the crushed man of Lee Mill Cave to the spaceman of the Iron Range, and more.

In addition to the precautions noted for various underground sites, a word is in order about safety generally. While all sites are publicly accessible and many are accessible to wheelchairs and strollers, several involve moderately difficult hikes and a few are very challenging. A light jacket and good walking shoes can keep you comfortable on guided underground tours. A helmet and several light sources can keep you safe on wild caving trips. Remember not to cave alone and to tell someone where you're going and when you'll return.

This book is only an introduction to Minnesota's underground and in-ground attractions. Learning about the state's geology and subterranean history can deepen your understanding and appreciation of its rivers, springs, lakes, and other natural features. *Minnesota Underground* can point out intriguing aboveground features and increase your knowledge of the state's down-under world. If, after exploring the sites listed here, you want to go to the next level, it's time to connect with the Minnesota Speleological Survey, Minnesota Caving Club, geology classes, and other organized groups that can safely guide you further into Minnesota's underground world.

A word about preservation is in order, too. When you visit Minnesota's subterranean attractions, you will see signs of human impact, both positive and negative. Historically, mining has both cost lives and spurred economic development. Reclamation efforts have converted former pit mines into beautiful lakes used for scuba diving and fishing. Agriculture has both benefitted the state economy and sometimes harmed its environment through agricultural runoff and the use of sinkholes as waste dumps. Visitors have both shared wondrous photos of Minnesota's underground and marred it with trash and graffiti.

Your choices and actions make a difference in the state's ecology. Do you toss a cigarette butt onto a cave floor, or do you pick up a soda can along a hiking trail? Do you pocket a prize fossil picked up from an area where fossil hunting is prohibited? Or, do you point it out to your companion, photograph it, and leave it where you found it? Especially if you travel under the watchful eye of children, your actions may have a long -term impact on Minnesota's environment.

Taking care of the state's subterranean world is another way to enjoy and appreciate it. This guidebook invites you to take a deeper view of Minnesota's lands, how they formed over the millennia, and how we humans have continued to shape and change their contours. The underground and in-ground attractions described here can help chart a course for an adventure or a unique family excursion. Make your travels count: make them memorable, unordinary, and underground.

WELCOME TO

MINNESOTA UNDERGROUND

KARST COUNTRY

The word "karst" is thousands of years old, referring to a stony place, associated with the landscape of southeastern Europe. But it was not until 1944 that "a karst type of topography" was described for Minnesota. The word did not seem to stick, however, until the 1970s.

A karst landscape often features sinkholes, caves, underground streams, and springs. Streams flowing into karst areas from elsewhere may end abruptly in blind valleys, continuing to flow downstream in subterranean channels, finally exiting as springs.

This network of underground channels with fast-flowing water marks karst country, even when the aboveground topography may not always reveal its presence. Karst topography may be obvious aboveground only where glacial drift is less than 50 feet thick or so, for instance, in southeastern Minnesota.

The Department of Earth Sciences at the University of Minnesota began mapping karst features in the 1970s, and the Department of Natural Resources is now responsible for mapping karst features like springs and sinkholes. University of Minnesota researcher Calvin Alexander has contributed substantially to this effort and generally to the understanding of karst in Minnesota and surrounding states.

SOUTHEASTERN MINNESOTA

Southeastern Minnesota contains the heart of the state's cave country, with more than 300 natural caves in Fillmore County alone. This is the part of the state characterized as "karst," a naturalized foreign term borrowed from southeastern Europe. Karst is a landscape dominated by soluble rocks like limestone. Instead of surface drainage, water enters sinkholes, flows through caves, and exits as springs. Consequently, there are few lakes or streams. Most of Minnesota's karst shows a mixture of these traits, however, and is therefore sometimes called "fluviokarst" by purists. In any case, there's much to see. Enjoy!

NIAGARA CAVE
FILLMORE COUNTY

The namesake subterranean waterfall in Niagara Cave on the Minnesota-Iowa border. (Postcard from the Greg Brick Collection)

Approaching the cave entrance building from the parking lot, the first thing you notice is a slanting trough of running water. Simulated "Sandy Creek" is a sluice where children and adults can realize their dreams of panning for gemstones. You can purchase a bag of mining "rough," place it in a screen-bottomed wooden box, dip it in the sluice, and watch your gems appear as the soil washes away.

The real gem, however, is beneath your feet. Niagara Cave plunges underground along a series of fractures in a cross-section pattern, with "section" lines at approximate 45-degree angles. The cave has a length of 1,750 feet and the greatest vertical relief of any Minnesota cave—150 feet. The tour traverses most of this length, following a zigzag path through the eroded fractures. In some passages, the ceiling soars far above your head; in other places, you walk over a grate and can see the cave floor scores of feet below. As the tour snakes downward through narrow gorges and canyons—some no more than four feet wide—you may feel as if you are walking through the eye of a very large needle.

Descending, you soon hear a muffled thunder from somewhere ahead. Your guide stops to describe the 60-foot waterfall ahead, since it would be useless for him to try to describe the feature when you finally stand looking over the precipice in the midst of spray and a crescendo of cascading water. Sixty feet below, the water churns in chaos; 70 feet above, the domed ceiling shelters the echoing tumult.

The waterfall and a wishing well are down the opposite arms of a short side path intersecting the main passage. In addition to the standard cave wishing well (this one is water-filled pothole scoured out by a whirlpool), Niagara provides another make-your-wish opportunity: The legend of Kissing Rock (a limestone shelf that fell from the ceiling) is that if you kiss someone on this table, the person you kissed would be yours forever.

As Niagara Falls is a traditional honeymoon destination, so romance abounds at namesake Niagara Cave. More than 400 couples have been married in the Crystal Chapel, which features a white lectern, benches, and archway, as well as natural decorations. Soda straw stalactites hang from the ceiling in the rear of the chapel; and "stained glass" windows, ribbon stalactites (cave bacon), and flowstone decorate the walls. A bit farther along the tour, is another romantic favorite: The Bridal Veil is a 75-foot tall cascade of ribbons and flowstone.

Every visitor has at least one favorite formation. Other candidates:

- The Wedding Cake has three layers and flowstone frosting.

- A skeletal hand and arm—complete with radius and ulna—protrude from high on the wall near the Kissing Rock.

- The Stalactite Room features numerous huge limestone icicles, thousands of years old, including sharks' teeth stalactites, which drop down along the wall.

- The Liberty Bell—complete with crack—was formed by an ancient whirlpool.

- Paul Bunyan's huge bed was eroded by rushing water.

- The Elephant's Head is composed of dried-up flowstone and is no longer growing.

- The Battleship is an aptly named lump of limestone.

In many places throughout the cave, you can find various fossils. Several examples of *Fisherites*, believed to be a form of algae, exist high on the cave walls. These once lived in the ancient seas that covered southeastern Minnesota. A cross-section of one has the appearance of a sunflower. A cephalopod visible near eye level also catches your attention.

Here and there, lumps of chert stick out of the wall, where other rock has eroded around the hard substance. Niagara Cave plunges through three distinct limestone layers: The Dubuque Formation lies nearest to the surface. Underneath it lie the Stewartville and the Prosser Members of the Galena Formation. All three layers were formed when seas covered this area 450 million years ago.

At the deepest point in the tour, the Stalactite Room, you are more than 200 feet beneath the surface. Many more rooms are accessible only with scuba gear. The main branch of the river that runs through the cave comes to the surface three miles to the south at Hawkeye Spring in Iowa, which runs into a tributary of the Upper Iowa River.

The current owner of Niagara Cave, Mark Bishop, continues to clear the lower passages. People have been digging out and exploring the cave since it was discovered in 1924 by two farm boys, who followed three lost pigs into a hole in the earth. The boys heard the pigs squealing and followed them to a ledge 75 feet below the surface. After the boys used a rope to rescue the pigs, they returned to explore the cave. More than three quarters of a century later, adventurers continue to visit Niagara's treasures.

One feature modern-day visitors don't see is a bat of any type. Because the entrance building covers the only opening to the cave, bats are unable to gain admittance to Niagara.

In 2015 Niagara Cave became the first commercial cave in the world to use solar energy to fully meet all its energy requirements. A 210-foot long photo-voltaic solar panel array produces 45,000 Kilowatts per year.

Directions: From Rochester, follow U.S. 52 to Harmony. In town, take State Highway 139 south and drive 2.5 miles to County Highway 30. Turn west and proceed 2.5 miles to the cave. Watch for signs. Almost at the Minnesota-Iowa border, Niagara Cave is less than an hour drive from Rochester and from exits off I-90 in Austin and Winona.
(*Continued on next page*)

Seasons/ Hours:	Open weekends in April, tours depart at 11 a.m., 12:30 p.m., 2 p.m., and 3:30 p.m. Check in 15 minutes prior to tour departure. May 1 to Memorial Day weekend, open daily; tours depart at 11 a.m., 12:30 p.m., 2 p.m., and 3:30 p.m. Memorial Day weekend through Labor Day weekend, open daily 9:30 a.m. to 5:30 p.m.; tours depart every 15 to 20 minutes. 9:30 a.m. to 5:30 p.m. After Labor Day through the last weekend in October, open daily, 10 a.m. to 4:30 p.m.; tours depart at 11 a.m., 12:30 p.m., 2 p.m., and 3:30 p.m. Fee. Group rates available on weekdays to schools, scouts, churches, and other organizations with a minimum of 15 admissions. Reservations must be made at least one week in advance. Private tours and weddings also available by reservation.
Precautions:	One-hour tour involves walking a mile plus descending and ascending 275 steps in humid environment; not advised for individuals with heart conditions, severe asthma, or similarly limiting issues. Temperature is 48 degrees year-round. A light jacket and walking shoes are recommended—no flip-flops, Crocs, high heels, or bare feet. Because of the stairs, wheelchairs and strollers cannot be used on the tour. Backpacks and backpack baby carriers are not allowed, but front packs for babies are fine. Pets are not allowed in the cave or gift shop. Certified service dogs are permitted; however, emotional support and therapy animals are not allowed.
Amenities:	Picnic tables, playground equipment, gemstone mining, mini golf, restrooms, gift shop, and concession stand open Memorial Day weekend through Labor Day weekend.
Contact:	Niagara Cave, P.O. Box 444, Harmony MN 55939. Ph: 507-886-6606 or 800-837-6606. E-mail: niagaracave@gmail.com. Web site: www.niagaracave.com.

MYSTERY CAVE I
FILLMORE COUNTY

The heart of Minnesota's karst country, Mystery Cave today is a maze cave system, comprising Mystery I, Mystery II, and Mystery III caves, with additional passages no doubt waiting to be found. Discovered first, Mystery I lies adjacent to and generally southwest of Mystery II, which lies southeast of Mystery III. With about 13 miles of known passages, the Mystery Cave system is second in length only to Cold Water Cave in Iowa among Upper Mississippi Valley caves. The cave system lies in a satellite parcel of the Forestville/Mystery State Park.

Mystery Cave offers a superb introduction to the geology of southeastern Minnesota. It features a network of linear passages in three predominantly limestone layers (the topmost Maquoketa Formation, the Dubuque Formation, and the lowest Stewartville Member of the Galena Formation). These three layers are themselves each comprised of many different identifiable levels. All three layers formed during the Ordovician Period (485 to 444 million years ago). This was, of course, long before the advent of mammals, never mind humans. Vast seas teemed with squid-like creatures

THE DRIFTLESS AREA

The Driftless Area, a designation popular in modern media, is obsolete scientific terminology for a 10,000-square-mile area in four contiguous states: extreme southeastern Minnesota, southwestern Wisconsin, northeastern Iowa, and northwestern Illinois. By far the greatest portion is in Wisconsin.

To understand what "driftless" means you must first understand what "drift" is. Drift refers to glacial deposits such as sand, gravel, and till. But before 1840, when the glacier theory was popularized, the origin of these deposits was a mystery. One suggestion was that the interior of the continent was covered by a lake or sea. Supposed icebergs from the polar regions would then explain the glacial deposits. As the drifting icebergs melted they dropped the materials we now think of as glacial deposits. Hence the name drift. But even when the glacier theory finally won out over the iceberg theory, the old name stuck.

Early geologists traveling through the Midwest noted something peculiar, however. This "drift" seemed to be absent from the area delimited above. Josiah D. Whitney was the first geologist to actually map out the area and call it "driftless" in 1862.

Why did glaciers avoid the Driftless Area? Paradoxically, the Driftless Area is actually a low spot relative to surrounding areas, so it should have easily filled up with ice. But according to one long-held explanation, Wisconsin's northern highlands deflected the ice stream. Another argues that this area is driftless because of its large swaths of porous sandstone. The subglacial water on which the glaciers slid drained away into the sandstone, grounding the ice front.

Fast forward to 1960. Geologist Bob Black discovered what he interpreted as glacial phenomena in the Driftless Area. What we now realize is that while the Driftless Area was not glaciated during the latest ice age, glaciers might have covered it during earlier ice ages, but the "drift" from those is so old that it has disintegrated, leaving few traces. We must at least acknowledge that the southeastern Minnesota karst, sometimes erroneously included in the Driftless Area, has a mantle of drift, which thins out toward the east.

and other marine life; primitive plants emerged; sediments and fossils built up on ocean floors. Mystery Cave visibly documents this period with its fossils of brachiopods, cephalopods, and other early sea creatures embedded in the ancient limestone.

Many millions of years later—still many years before recorded human history— the solution cave system began forming along joints in the rock layers as weak acidic rainwater seeped downward, dissolving the limestone. The Root River facilitated cave development, as its South Branch took a shortcut underground, beneath a surface river meander. The underground stream, which resurfaces at Seven Springs (see SEVEN SPRINGS WILDLIFE MANAGEMENT AREA), can be seen in the deeper levels of Mystery Cave, not open to tourists. The South Branch Root River entirely sinks into the cave during dry years.

Mere thousands of years ago, stalactites, stalagmites, flowstone, and other multi-hued formations began forming as groundwater containing calcium carbonate trickled into the air-filled cavities and precipitated. The release of carbon dioxide into the cave air caused the calcium carbonate in the groundwater to precipitate and the formations (or "speleothems") grew. Many are dripping and still growing today.

Caves also developed in southeastern Minnesota thanks to the region's relative lack of glaciation. While deep debris, or glacial drift, left from melting and retreating glaciers 10,000-plus years ago covers much of the state, the drift here is less common and less deep. Of four major glacial advances over the last million years, only the first two covered this "Driftless Area."

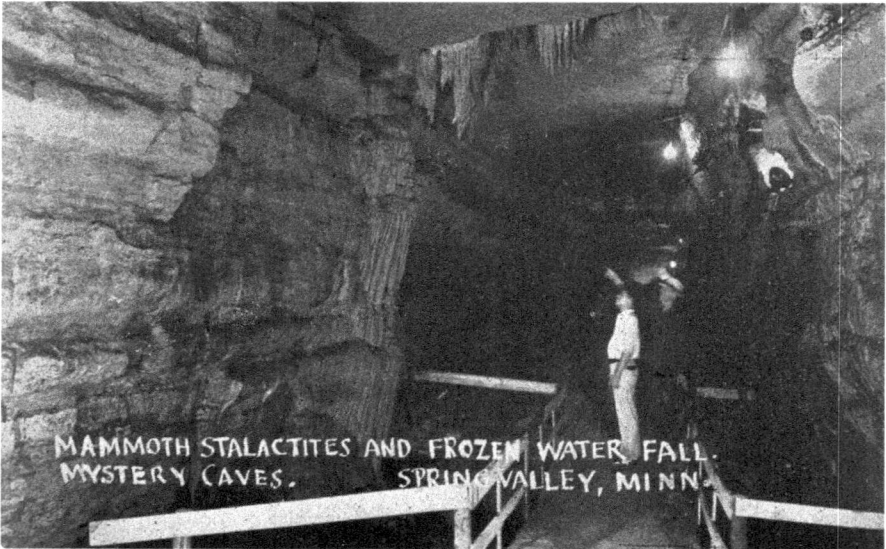

Mystery Cave 1—Historic image depicts delights of early Mystery Cave tours. (Postcard from the Greg Brick Collection)

Mystery Cave has been the focus of more scientific studies than other Minnesota caves. This is where the first really detailed Minnesota cave sediment studies took place. From radiometric dating of flowstone, we know the cave is more than 160,000 years old—since before the latest glacial advance. Being near the margins of the continental ice sheet, glacial sands and gravels were washed into the cave, as seen best at Enigma Pit.

The existence of Mystery Cave was first surmised by state geologist N.H. Winchell in 1876, based on where water entered the ground, and where it came back out again at springs. But in the absence of caving clubs, these clues were not followed up.

More than 60 years later, Joseph Petty discovered Mystery I (then called Mystery Cave) in 1937, after observing melted snow and steam near crevices in the rock surrounding the entrance to an underground stream. He explored the cave with property owner Walter Blakeslee, who in 1938 began offering a one-hour tour similar to the tour offered today. Mystery I was closed during World War II. After the war, new owner Clarence Prohaska improved the cave and again opened it for tours in 1948.

Today the one-hour, stroller- and wheelchair-accessible Scenic Tour of Mystery I begins at a large steel door anchored in the side of the bluff rising above the South Branch of the Root River. Tiny windows in the door permit entry and egress to the several species of bats that hibernate in the cave. The entrance is in the Dubuque Formation, featuring olive gray limestone with interbedded shale. The northeast trending passage is an easy walk: It's paved, well-lit, 10 to 20 feet wide, and up to 30 feet in height.

The Mystery I tour offers the following memorable features:

- The Cathedral Room features a chapel with a 30-foot-high vaulted ceiling reaching up to the Maquoketa Formation, another distinct limestone layer.

- The Stalactite Room offers a variety of other slowly growing formations, in addition to the rock icicles hanging from the ceiling.

- Turquoise Lake is a clear pool 10 feet deep with a flowstone bottom, stalactites, and cave popcorn concretions along one side.

- The Bomb Shelter, once designated as a wartime bomb shelter, is the final room of this tour.

In addition to both natural and artificial pools, the Mystery Cave system contains several streams, including the subterranean section of the South Branch of the Root River.

Cooled by its sojourn underground, the river downstream of the cave supports a healthy trout population; and you can find good fishing in Forestville State Park.

Directions:	From Rochester, follow U.S. 63 to State Highway 16. Turn left on Highway 16 and drive west through Spring Valley. Turn right on County Highway 5 and drive four miles to County Highway 118. Follow County Highway 118 and watch for signs to the cave. Mystery Cave is less than an hour drive from Rochester and from exits off I-90 in Austin and Winona.
Seasons/ Hours:	Open weekends from mid-April to Memorial Day weekend and after Labor Day through October; open daily from Memorial Day weekend through Labor Day weekend. Fee. A state park vehicle permit is also required.
Length:	The three-quarter-mile Scenic Tour lasts about one hour.
Precautions:	Featuring paved passages and metal bridges, the Scenic Tour is accessible to strollers and wheelchairs. A light jacket is recommended; temperature is 48 degrees year-round.
Amenities:	Forestville/Mystery Cave State Park offers picnic tables, playground equipment, restrooms, trout streams, and recreational trails. Also in the park, Historic Forestville is a restored 1850s village operated by the Minnesota Historical Society. If you visit in May, watch for the carpet of Virginia bluebells along the drive to the park office.
Information:	Forestville/Mystery Cave State Park, Route 2, Box 128, Preston, MN 55965. Mystery Cave phone: 507-937-3251. Website: www.dnr.state.mn.us/state_parks/index.html and use the Park Finder.

BATS AND WHITE-NOSE SYNDROME

One of the most important bat hibernacula in the Upper Midwest, Mystery Cave supports more hibernating bats than any other natural cave in the state. The numbers of hibernating bats, however, have been in steep decline in recent years, due to the spread of a fungus, *Pseudogymnoascus destructans* (abbreviated "Pd"), which causes "white nose syndrome" (WNS). A February 2019 count in Bat River Cave, for example, showed a 98 percent decline in its bat populations since 2011.

The WNS name comes from the white fuzzy growth observed on the nose, ears, or wings of infected bats. The fungus, which may irritate bats' skin, disrupts their hibernation, causing them to expend energy, dehydrate, and starve. Infected bats often fly out in the middle of winter, in a fruitless search for food. The fungus thrives in humid, cool temperatures in the 40- to 55-degree range, just like the environment of many Minnesota caves and mines.

BATS AND WHITE-NOSE SYNDROME (*CONTINUED*)

Minnesota provides a home to eight species of bats, with four species migrating south for the winter and four hibernating in the state. The species that hibernate (little brown bat, big brown bat, northern long-eared bat, and tricolored bat) are more vulnerable to WNS. All four are listed as Special Concern Species in Minnesota and the northern long-eared bat is listed as a federally Threatened Species, thanks largely to WNS. Minnesota's bats are quite small, weighing from two-tenths of an ounce to just over an ounce.

WNS was first reported in eastern New York during the winter of 2006/2007 and five years later was detected in northeastern Minnesota at the Lake Vermilion–Soudan Underground Mine State Park. Spreading rapidly, WNS has now been confirmed or suspected in 32 states. Millions of bats have perished. In the winter of 2017/2018, WNS had been found in 13 Minnesota counties and is likely present in all counties where bats hibernate.

One positive impact of the disease, however, has been increased awareness of the importance of bats in the ecosystem, although misunderstanding still abounds. Some people still fear bats, which can carry rabies, and stories about large, tropical vampire bats feeding on the blood of livestock fuel misunderstanding of all bats. Yet, bats consume millions of bugs and reduce insect populations that harm agriculture and annoy and infect people. Bats catch mosquitoes on the fly, in their cupped tail membranes and then transfer the bugs to their mouth. Some bats aid in plant pollination, seed dispersal, and scientific advancement. Bat research has improved understanding of hearing, sonar, and blood coagulation, among other advances.

Despite the devastation of WNS, there has been some recent good news in the battle against the fungus. In Europe, from where the disease originally spread to the United States, all bats have natural immunity to the fungus. There is evidence that some bats in the United States survive the disease. And some researchers are seeking possible remedies.

A team of scientists from the U.S. Forest Service, U.S. Department of Agriculture and the University of New Hampshire published their study in the journal Nature Communications, writing, "*P. destructans* is unable to repair DNA damage caused by UV light, which could lead to novel treatments for the disease." Only a few seconds of moderate exposure from a hand-held UV-C light source killed 99 percent of *P. destructans*.

BATS AND WHITE-NOSE SYNDROME (*CONTINUED*)

At the Center for Drug Design in the University of Minnesota's School of Pharmacy, Associate Professor Christine Salomon leads a research group that studies the chemistry of bacteria and fungi in order to develop treatments for diseases like WNS. The team collects samples from field locations, isolating microbes, growing them and then testing the compounds that these cultures produce.

The team collected more than 2000 different bacteria and fungi at various sites—including Soudan Mine, Mystery Cave, and several Twin Cities brewery caves—and tested each strain for the ability to kill the bat pathogen in the lab. The team identified more than 100 different bacteria and fungi that can kill the bat pathogen. The researchers are now testing these strains against the fungus on natural surfaces relevant to hibernacula.

The team is also studying the fungus itself. Since the disease is so new to science, there are still many unanswered questions, such as, "How long can the fungus survive at different temperatures?" or "Where is the fungus concentrated in hiberernacula?" Answering these questions will also help develop treatments for WNS.

Although the team is likely more than a year away from developing a safe and potentially effective treatment, Salomon's "biological control" research is stupendous news for bats. Eventually the team will need to test potential treatments in caves or mines.

While the DNR is the lead agency for WNS response in Minnesota, it works with other state, federal, provincial and tribal agencies, national health laboratories, academic researchers, and other organizations to monitor the spread of WNS and better understand and control the disease. (See: www.dnr.state.mn.us/wns/index.html.)

Federally, the U.S. Fish and Wildlife Service leads the U.S. response in combating the disease and is a partner in www.whitenosesyndrome.org, which brings together research and recommendations from all sectors. Other agencies involved include the National Park Service and U.S. Forest Service. Several universities, the National Speleological Society, and National Caves Association are among scores of non-federal partners.

WNS primarily spreads from bat-to-bat contact, but the fungus can also be carried from one location to another on clothing and equipment. The DNR is working to control the spread through education and prevention. Public tours of Soudan Underground Mine and Mystery Cave feature a brief

BATS AND WHITE-NOSE SYNDROME (*CONTINUED*)

presentation on how to prevent the spread of WNS. Before and after the tours, visitors walk across special mats designed to remove spores from footwear.

Other steps you can take to ensure bats' long-term survival:

- If you find dead, sick, or injured bats contact the DNR: www.dnr.state.mn.us/reportbats/index.html or local U.S. Fish and Wildlife Service Field Office.

- Do not enter caves or mines that have hibernating bats. One researcher says, "There's a season for caving and a season for bats."

- Follow recommended decontamination procedures (www.whitenosesyndrome.org) when visiting mines and caves used by bats.

- Enhance bat habitat on your property by retaining large trees, protecting wetlands, and building bat houses.

Research regularly uncovers new knowledge about bats and WNS. You can help by staying informed and sharing what you learn.

Prof. Salomon collecting samples in Banholzer Cave during a WNS investigation. (2015).

MYSTERY CAVE II
FILLMORE COUNTY

A one-hour Lantern Tour of Mystery II offers visitors a clear lesson on the difference between limestone and dolomite. The tour descends through layers of limestone and shale into dolomite, which is rougher in appearance here. The shapes of the walking passages are also different in the two layers. Passages within the dolomite tend to have rounded or even pointed ceilings along vertical joint planes, while passage ceilings in the limestone layers tend to be as flat as a mine tunnel. Sometimes a limestone layer has fallen from the ceiling, where a softer layer of shale intersperses two limestone layers, creating the flat appearance.

A short walk from the entrance leads to a large, straight passage, Fifth Avenue. To the left (east) it leads to the Garden of the Gods, featuring numerous stalactites and other formations on ceiling, walls, and floor. To the right (west) it leads to a variety of rooms and side passages. While visitors will see many formations, in general, fewer speleothems exist in the long, drier passages of Mystery II, compared to Mystery I.

A few highlights of the Mystery II Lantern Tour:

- The Carrot Sticks are twin stalactites, colored by hydrous iron oxide incorporated into the calcium carbonate layers of the formation.

- The Hills of Rome feature beds of stone rising through the Stewartville and Dubuque layers.

- Blue Lake, the largest pool in Mystery, has a water level that fluctuates during the year, drying up completely each winter. The lake contains unusual raft cones and calcite rafts. When the water level drops, the rafts sink onto the cones and build them up. Some of the cones are seven feet high.

Mystery II (also known as Minnesota Caverns) was discovered in 1958, when Mike McDonald pushed through a narrowing passage in Mystery I. McDonald returned with a small group of cavers and further explored the long extension of this cave system. A new exit to the surface was dug, which still serves as the current Mystery II entrance. Mystery II is also accessible from Mystery I by negotiating a 10-hour walk, crawl, and squeeze— called the Door-to-Door Route.

When tours began at this second entrance to the system, the system was temporarily called "Minnesota Caverns." The tour followed passages in the Stewartville layer of the Galena Group and highlighted Fourth and Fifth Avenues and the Garden of the Gods—a creamy rich welter of flowstone.

Since then, additional passages and rooms have been discovered and mapped. In 1967, cavers Doug Nelson and Craig Porte discovered Mystery III; other cavers further explored the new section, including Lily Pad Lake, Eureka Avenue, and Dragon's Jaw Lake. Each year, cavers explore, clear, and survey more of the far reaches of the Mystery Cave system. The Department of Natural Resources acquired the cave system in 1988, refurbishing the old tour routes.

With advance registration, the DNR offers several additional tours, for instance, a two-hour Geology Tour of Mystery II by flashlight provides added geological background. Do you know the definition of a "vug" (a small cavity lined with crystals, similar to a geode)? Or, would you recognize a "helictite" (a hook-, twig- or bush-like formation created when water seeps slowly out of the side of a stalactite instead of dripping straight off the end)? Helictites tend to form in drier caves.

Photography Tours allow photographers to enter either Mystery I or Mystery II and set up their tripods and cameras for perfect subterranean shots. School Tours and Advanced Educational Tours are aimed at students through college level. Four-hour Wild Caving Tours involve crawling and squeezing through narrow, undeveloped passages.

Directions, season/hours, amenities, and contact information are the same as for Mystery I.

Directions:	From Rochester, follow U.S. 63 to State Highway 16. Turn left on Highway 16 and drive west through Spring Valley. Turn right on County Highway 5 and drive 4 miles to County Highway 118. Follow County Highway 118 and watch for signs to the cave. Mystery Cave is less than an hour drive from Rochester and from exits off I-90 in Austin and Winona.
Seasons/Hours:	Open weekends from mid-April to Memorial Day weekend and after Labor Day through October; open daily from Memorial Day weekend through Labor Day weekend. Fee. A state park vehicle permit is also required.
Length:	The Lantern Tour takes one hour.
Precautions:	Expect to carry your own lantern/flashlight. The Mystery II trail is occasionally wet and slippery. This tour is more physically challenging than the Scenic Tour and is not available to children under eight. No sandals or open-toed shoes are allowed. You will be on rugged gravel paths and uneven stone stairs. The Wild Caving Tour is offered only to people age 13 and older, who are in good physical condition. Boots and gloves required.
Amenities:	Forestville/Mystery Cave State Park offers picnic tables, playground equipment, restrooms, trout streams, and recreational trails. Also in the park, Historic Forestville is a restored 1850s village operated by the Minnesota Historical Society. If you visit in May, watch for the carpet of Virginia bluebells along the drive to the park office.
Information:	Forestville/Mystery Cave State Park, Route 2, Box 128, Preston, MN 55965. Mystery Cave phone: 507-937-3251. Website: www.dnr.state.mn.us/state_parks/index.html and use the Park Finder.

SEVEN SPRINGS WILDLIFE MANAGEMENT AREA
FILLMORE COUNTY

Seven Springs, the big springs that drain Mystery Cave, were locally so well known that the area appeared on postcards. Although called "Seven Springs," there are many more in the vicinity. The newly established Seven Springs Wildlife Management Area offers public access to them. The Department of Natural Resources manages this approximate 275-acre parcel primarily for deer, turkey, small game, and pheasants, as well as timber harvests. Visitors to the springs need to be prepared for a challenging hike. Native grasses are planted on the ridge tops, with oak and other hardwoods on the slopes and floodplain.

Directions:	From Wykoff follow County Road 5 for four miles south, turn right onto County Road 12 and follow it for 1.25 miles. Then turn left on Orion Road and drive about 1.5 mile to the parking area next to (and north of) the river.
Seasons/Hours:	Open year-round.
Length:	The hike from the anglers' parking lot to Seven Springs area is approximately 1.5 miles along the river banks.
Precautions:	Challenging access from parking area. Avoid hunting seasons or wear blaze orange/pink.
Amenities:	There are no facilities.
Information:	Phone: 507-206-2858. Website: www.dnr.state.mn.us/wmas/index.html and search for Seven Springs.

CANFIELD SPRING CAVE, FORESTVILLE STATE PARK
FILLMORE COUNTY

The main section of Forestville/Mystery Cave State Park also contains a few small caves, though they are off the recreational trails and difficult to find. Big Spring on Canfield Creek, however, is worth the hour-plus trek to the end of a long spit of land that juts south from the rest of the park property.

Water from the spring has dissolved out a small cave in the side of a bluff and in summer and fall spreads out across a pebble "beach" before entering the main current of the creek. Wearing boots, you can stand at the entrance of this cave, as the water flows over your feet. Though it may appear to issue from the depths of the earth, this stream actually sank underground less than five miles upstream.

Boots make a hike to the spring easier, since it involves two fords across the creek. Of course, you could solve this problem by riding horseback to the spring. There is a hitching post nearby.

THE MINNESOTA SPRING INVENTORY

Minnesota famously has 10,000 lakes but it has even more springs. Springs were far better known a century ago than they are today, because they were an important water supply. Camp Coldwater, in Minneapolis, supplied drinking water to the soldiers who built Fort Snelling, beginning in 1819. Today, however, drinking from an untested spring is never recommended, due to various pollutants—from agricultural runoff to old trash discarded in convenient sinkholes.

In southeastern Minnesota, most springs issue from limestone and many are the discharge points for cave systems. An example is found at Seven Springs Wildlife Management Area, where the South Branch of the Root River discharges from Mystery Cave. Odessa Spring, the largest in the state, is located on private land on the Upper Iowa River. Odessa drains from a suspected cavern many miles in length.

Minnesota's greatest spring hunter was Thaddeus Surber (1871 – 1949), an aquatic biologist who mapped the springs of the Root River, of Pine County, and the North Shore of Lake Superior. The latter day equivalent was the DNR's Minnesota Spring Inventory (2014 – 2018) which mapped several thousand more springs. For convenience, the state was divided up into seven "crenoregions" (*crenos*=spring) of which the southeastern karst region is just one example. Most of the other springs are glacio-fluvial in origin. Several historic mineral-water spas were rediscovered during this project.

Watch for wildlife along the wooded trail. During one fall visit, a pileated woodpecker flew overhead and three bald eagles soared above the spring and bluffs. You may also spot turkeys, soaring turkey vultures, and scarlet tanagers.

Deer, beaver, raccoon, mink, red and grey fox, and other animals make their homes in Forestville State Park, as do a growing number of coyotes. If you begin your hike to Big Spring later in the day, when you return at dusk you may hear scores of them, yipping and howling and crying from one bluff to another, their voices amplified in the resounding valley. Timber rattlesnakes have been spotted but are of very little threat if left alone.

The park namesake is the town site of Forestville, founded in 1853, as a trade center where farmers exchanged their produce for whatever they needed. Once boasting two general stores, a grist mill, a brickyard, two hotels, a school, and a population of 100, the village declined after the first area railroad bypassed it in 1868.

There are many other springs in the vicinity (see SEVEN SPRINGS WILDLIFE MANAGEMENT AREA). Rainy Day Spring is in Forestville State Park, though there isn't a trail to it. Moth Spring, outside the park at the head of Forestville Creek, is accessed through the Maple Springs Campground, where you can pay to park your vehicle while you hike. The drive to this private campground is near the entrance to the state park.

Directions:	From Rochester, follow U.S. 63 to State Highway 16. Turn left on Highway 16 and drive west through Spring Valley. Turn right on County Highway 5 and drive four miles to County 118. Follow County Highway 118 about 2 miles to the park. Forestville State Park is less than an hour drive from Rochester and from exits off I-90 in Austin and Winona.
Seasons/Hours:	Open year-round, however, some facilities (shower building, campgrounds with electricity, campgrounds for horses) and trails close in the winter. Fee.
Length:	The walk from the anglers' parking lot to Big Spring is 3.5 miles.
Precautions:	Big Spring is inaccessible during the springtime and other periods of high water. Since the spring is located on public hunting land, wear blaze orange/pink during hunting season. The trail to the spring includes two fords over Canfield Creek—easy on horseback, somewhat challenging on foot. Waterproof footgear is recommended.
Amenities:	Forestville/Mystery Cave State Park offers picnic tables, playground equipment, restrooms, trout streams, and recreational trails. Also in the park, Historic Forestville is a restored 1850s village operated by the Minnesota Historical Society.
Information:	Forestville/Mystery Cave State Park, 21071 County 118, Preston, MN 55965. Mystery Cave phone: 507-937-3251. Website: www.dnr.state.mn.us/state_parks/index.html and use the Park Finder.

CHERRY GROVE BLIND VALLEY SCIENTIFIC AND NATURAL AREA
FILLMORE COUNTY

A few miles south of Mystery Cave you can walk above ground over the headwaters of another cave system. Donated to the state by the Thomas Kapper family, the Cherry Grove Blind Valley Scientific and Natural Area preserves a 40-acre parcel of karst from all development. The site is immediately south of Cherry Grove Wildlife Management Area and its planted native grasses.

The Department of Natural Resources has gated the entrances to several caves, including Goliath's Cave and Woodchuck Cave, for safety. (Some passages are largely water-filled, and much of this cave system is prone to flooding during thunderstorms.) But you can still walk above ground and get a close-up look at sinkholes and a sinking stream at the end of a blind valley.

Jessie's Kill flows into the site from the southwest, eventually disappearing in a blind valley near Downwater Sinkhole. To the northeast, Woebegone Sinkhole marks the path of the now underground stream.

Originally called Coon Cave, Goliath Cave's entrance opens at the base of a depression, with walls showcasing layers of limestone and shale. The cave system runs eastward beyond the SNA boundaries. The adjacent land is owned by the Minnesota Cave Preserve, which drilled a new entrance to Goliath Cave in 2004. The story of the owner's battle with the DNR for ownership of the cave is the subject of Cary Griffith's 2009 book, *Opening Goliath*.

Farther east, water from the Goliath Cave system re-emerges at Big Spring in Forestville State Park and enters Canfield Creek. Using dye to trace the water's paths, researchers documented this connection and continue to study the hydrology of the cave and region. To learn more, visit: https://conservancy.umn.edu and search for Hydrology of Goliath's Cave.

Directions:	From Rochester, follow U.S. 63 to State Highway 16. Turn left on Highway 16 and drive west through Spring Valley. Turn right on County Highway 5 and drive to the junction of County Highways 5 and 14 (1.0 mile south of the Mystery Cave turnoff on 5). Proceed 1.0 mile south on 5 to Cherry Grove. Continue south on 5 for 0.5 mile. Turn left (east) on County Highway 20 and drive 0.1 mile north on Township Road 379. The scientific and natural area is on your left (west). Turn left and drive .375 mile to the parking lot, which is on your left. There is an interpretive sign at the parking area. Lat/Lon: 43.581912, -92.264184.
Seasons/Hours:	Open year-round, dawn to dusk.
Length:	N/A
Precautions:	Closed below ground. Wear blaze orange/pink during hunting seasons.
Amenities:	Hiking, snowshoeing, cross-country skiing.
Information:	Scientific and Natural Areas Program, Minnesota Department of Natural Resources, 500 Lafayette Rd., Box 25, St. Paul, MN 55155. Phone: 651-259-5800. Website: https://www.dnr.state.mn.us/snas/index.html and use the A-Z list to search for Cherry Grove.

MINNESOTA CAVE PRESERVE
FILLMORE COUNTY

The five-and-a-half-mile long Spring Valley Caverns, a former show cave (late 1960s) is now the show piece of Ackerman's Cave Farm. John Ackerman has harvested a crop of three dozen caves on this land since about 1990, often using a trackhoe. The Cave Farm, in turn, is part of his Minnesota Cave Preserve, which includes caves scattered across Fillmore County, along with access acreage and underground rights to explore Cold Water Cave over the border in Iowa. With a surveyed length of over 17 miles, this cave system is the 32nd longest in the United States and has been declared a National Natural Landmark.

Goliath's Cave, with a surveyed length of 2.39 miles, also lies partially within the preserve. Its natural entrance, which has been gated, is located in Minnesota's Cherry Grove Blind Valley Scientific and Natural Area. In 2004 the preserve acquired several surface acres above the cave and created a second entrance. The upper level of the cave contains large dry passages, in contrast to the lower stream passages.

Altogether, the Minnesota Cave Preserve consists of seven separate preserves in southeastern Minnesota and northern Iowa. These properties encompass 42 caves, 714.3 surface acres, and 1,274 acres of additional subterranean cave rights. In August 2019, a new parcel was added to this list: Hiawatha Caverns is a highly decorated former show cave in Winona County.

Spring Valley Caverns and other caves within the preserve begin in their upper layer within the smooth Dubuque Formation and extend downward into the rougher Stewartville limestone. Most passages are 50 to 65 feet beneath the surface, though some are as deep as 135 feet. Many of the caves feature stalactites, stalagmites, flowstone, and other speleothems. The Sentinel, the largest and oldest cave column in Minnesota, stands in Spring Valley Caverns. Several of the caves contain streams at their lowest levels.

The Minnesota Cave Preserve is home to the Minnesota Caving Club, which, like the Minnesota Speleological Survey (https://sites.google.com/view/msscaves), is a grotto (member organization) of the National Speleological Society. Ackerman's goal in establishing the nonprofit preserve is to maintain the caves in their undeveloped state and make them available to cavers for education and exploration.

Seasons/ Hours:	Year-round, by appointment. Minnesota Caving Club members pay dues; however, some tours, for example, for scouts and school children, are free.
Length:	The preserve contains 36 miles of explored passages.
Precautions:	The caves contain no electric lighting or concrete walkways. Wear coveralls or a sweatshirt and plan to get muddy. Bring a change of clothing and three light sources. Temperature is 48 degrees Fahrenheit.
Information:	Website: www.cavepreserve.com/

Minnesota Cave Preserve—The Frozen Falls is one highlight of Spring Valley Caverns. (Photo: John Ackerman)

FILLMORE COUNTY HISTORY CENTER
FILLMORE COUNTY

The town of Fountain dubs itself the "Sinkhole Capital of the USA" but got the name "Fountain" from a big spring that you can still see flowing today along Keeper Road, one mile west of town. Once used as a town water supply, the spring water eventually became undrinkable owing to contamination from trash disposal in those very same sinkholes, but residents have made extensive efforts to improve the situation in recent years.

In fact, if you are traveling along US Highway 52, you will see the sign for the "Sinkhole Capital of the USA" as you pull into town at County Road 8. The sign is located in Roadside Park, and immediately behind it you will see a clump of trees. If you drive up the road a little bit and park, you will see that it's a fenced sinkhole with a viewing platform and signage.

Across the street at the Fillmore County History Center, you will find an exhibit, "Sinkhole Capital of the U.S.A.," which offers an "in depth" look at the karst landscape that has supported both agriculture and mining over the years. The exhibit explains sinkhole development and showcases items excavated from an old family trash site at the bottom of a sinkhole in Cherry Grove Blind Valley Scientific and Natural Area. Led by a university professor, a group of students restored the sinkhole and recovered a child's cap pistol, five-gallon salt glazed crock, jugs, jars, and glass medicine bottles, among other items. Many of the bottles were constipation remedies, telling you something about the pioneer diet. Elsewhere in the museum, the remnants of a velocipede, too large for the exhibit case, lean against a wall.

Another central hallway exhibit details efforts to mine iron ore inear the towns of Chatfield, Etna, Spring Valley, and Wykoff, to name a few. While prospectors sought iron deposits in the late 19th century, mining did not begin until 1942. By 1968, about eight million tons of low-grade ore were removed from more than 120 strip mines across western Fillmore County. The mines closed after the commercial development of taconite pellets proved to be a superior, cost-effective iron source. You can see examples of flooded mine pits at Goethite Wildlife Management Area, south of Spring Valley, Minnesota.

SCIMITAR CATS

Part of the Minnesota Cave Preserve, Tyson Spring Cave, near the town of Chatfield, appeared in stereopticon views and postcards from the early days, when it was a popular picnic spot. Spring water gushes from the entrance at 50 gallons per minute at the base of a 120-foot cliff of the Galena limestone. A commercialization attempt was made in the 1930s, but the cave was not thoroughly explored until the Argonaut Society and others came along with scuba gear in the 1970s, allowing explorers to get past the sump, or water-filled passages, to more walking passages beyond. Some of the ledges in the passages were lined with frogs, staring blankly at the determined explorers as they marched past. Ackerman installed a culvert to allow dry access beyond the sump to the estimated three miles of passages in this branch-work cave.

Most amazing of all were the scimitar cat bones found in Tyson Spring Cave in 2008. Scimitar cats are close relatives of saber-tooth cats. A paleontologist from the Illinois State Museum reported a radiocarbon age of 22,250 years for the bones, at which time the cat lived in a steppe-tundra environment and perhaps preyed on Ice Age mammoths. Surprising when you realize that the saber-toothed cat deposit most people are familiar with is the famous La Brea tar pits in faraway Los Angeles.

Strictly speaking, the complete name of the organization is the "Fillmore County History Center and Genealogy Library." The Emery & Almeda Eickhoff Genealogy Library is a destination for many genealogists because the county was one of the earliest settled in Minnesota. Many early settlers traveled to other regions via Fountain—"the Ellis Island of Minnesota," according to Debra Richardson, Fillmore County History Center executive director. The library's books, photographs, and microfilmed newspapers, as well as county birth, death, and marriage records can help family historians digging for details about their ancestors' lives. Of interest to cave and mine historians, the library also contains two archival boxes, one on sinkholes and one on iron mining.

The museum and library complex has occupied this site since receiving—and then restoring—a former elementary school, built in 1958. At recess, children would reportedly sled down the sides of a sinkhole on the property, since filled in.

How many sinkholes are there in Fillmore County? According to the Minnesota Karst Database, there are 10,910 mapped as this book goes to press. They're not uniformly distributed, but rather clustered into several sinkhole plains, the largest being the Fountain and the Harmony sinkhole plains. Paradoxically, Mystery Cave is not part of a sinkhole plain.

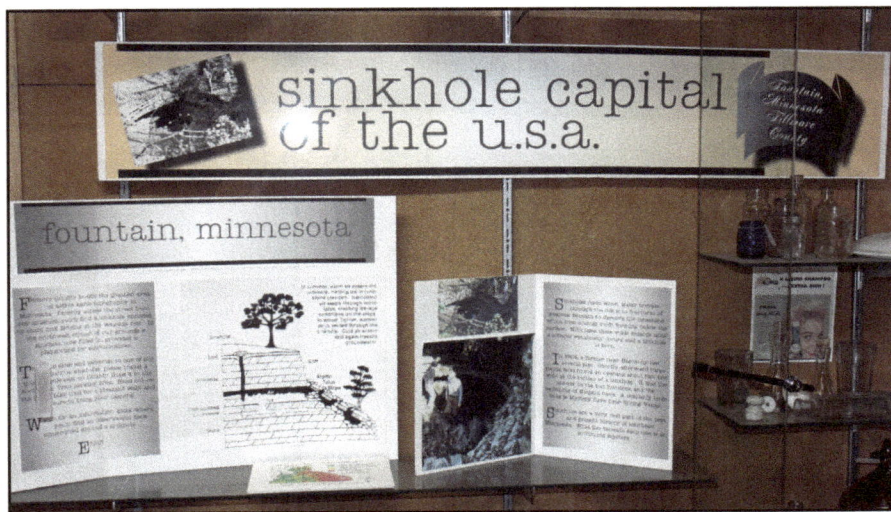

The Fillmore County History Center features a display on the "Sinkhole Capital of the U.S.A." (Photo: Doris Green)

Directions:	Located in Fountain at the intersection of U.S. 52 and County Road 8.
Seasons/ Hours:	The museum is open Tuesday – Saturday, 9 a.m. – 4 p.m. Make an appointment to see library materials. Fee.
Length:	N/A
Precautions:	N/A
Amenities:	The museum complex offers exhibits for folks with wide-ranging interests, from agricultural implements to replicas of a turn-of-the-century newspaper office and soda fountain, to toys and dolls. Two connected pole buildings, a country school house, and airplane hangar are also filled with artifacts and displays. Guaranteed not to bore any family member.
Information:	Fillmore County History Center, 202 County Rd 8, Fountain MN 55935. Phone: 651-259-5800. Website: https://fillmorecountyhistory.wordpress.com/.

ROOT RIVER STATE TRAIL SINKHOLE
FILLMORE COUNTY

Driving around southeastern Minnesota, you can't miss the tree islands marking the numerous sinkholes. Farmers avoid tilling near these depressions, which can extend into the dolomite and limestone bedrock. Yet driving around gives you only a distant view of these sinking dips in the landscape. For an up-close inspection of another sinkhole, head to the beginning of the Root River State Trail about a quarter mile east of the Fillmore County History Center and Roadside Park in Fountain.

Built on an abandoned railroad bed, the paved, 42-mile trail generally follows its namesake river all the way to Houston, from the trailhead in Fountain. Roughly a half mile farther east along the trail from the parking area, a large sign announces the sinkhole just north off the trail. A boardwalk and observation deck extending over the sinkhole provide a close-up lesson in geology, although the view is best in leaf-off seasons.

Detailed signage explains that sinkholes form when slightly acidic surface water percolates down fractures in soluble rocks like dolomite and limestone and enlarges them to form a cavity. Eventually the sediments above the cavity collapse into it, resulting in a sinkhole.

Hydrologists study water flow in a karst landscape by putting organic dye into sinkholes and tracing where and when the dye emerges in springs. Dye-trace studies show that water entering sinkholes near the trail is discharged from the springs northwest of Fountain. Water can travel miles per day underground, making the karst landscape vulnerable to pollution, since the moving water can carry pollutants far from their source. Scientists map underground water flow through channels and caves to its resurgence,

Sinkhole on private land. (Photo: Doris Green)

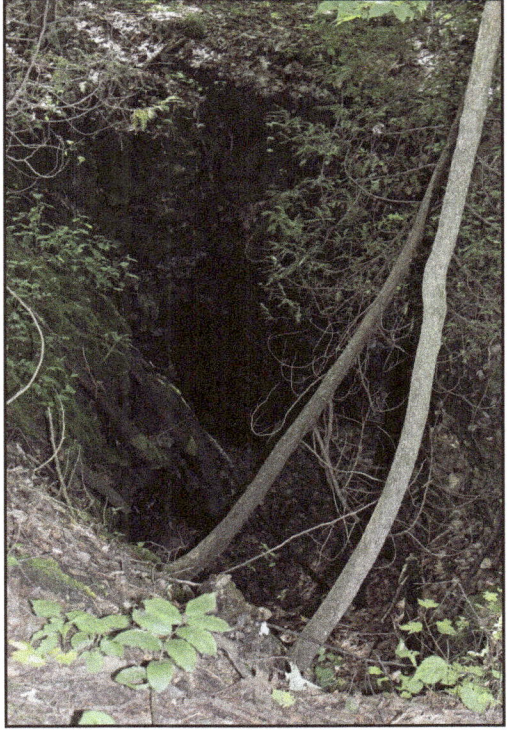

aiding pollution studies. Researchers also use LIDAR (light detection and ranging) technology to find sinkholes where pollutants may freely enter underground streams.

The berm around the trailside sinkhole helps prevent contamination of ground water from sediment-laden rainwater. A berm, dike, or grass plantings can slow erosion at the edge of a sinkhole. Sometimes retention ponds can be built beyond the edge of a sinkhole to further slow water flow into the sinkhole, while also filtering it through bedrock. Additional signage explains how individuals can lessen pollution in this karst landscape, for example, by not throwing trash into a sinkhole.

Directions:	The trailhead parking area is located on County Road 8 in Fountain about 0.25 mile east of the intersection with U.S. 52. The sinkhole is just off the trail about a half mile east of the trailhead.
Seasons/Hours:	Year-round.
Length:	N/A
Precautions:	Wheelchair accessible.
Amenities:	Boardwalk around the sinkhole, informational signage.
Information:	Website: www.dnr.state.mn.us/state_trails/index.html and use the Find by A-Z List to search for Root River.

GOETHITE WILDLIFE MANAGEMENT AREA
FILLMORE COUNTY

Goethite (usually pronounced "gur tite") is the mineralogist's name for the hydrous oxide of iron that was mined in southeastern Minnesota from 1942 to 1968. It was named in honor of the German poet Johann Wolfgang von Goethe (1749-1832), the author of *Faust*. While a respectable scientist in his own right, even dabbling in geology, Goethe's connection to the mineral is obscure.

Mining began during World War II when the nation was desperate for raw materials. The "brown ore," as it was called, was actually a mixture of goethite and hematite. One theory is that it formed by the weathering and concentration of iron compounds in the underlying limestone bedrock. More than 50 mines, some of them very small indeed, operated in three counties, shipping their ore to Missouri steel mills for processing. All of this, of course, is in stark contrast to the mining of the much older Precambrian iron ores in northern Minnesota. You can see several abandoned mining pits, now flooded and available for wildfowl, at this 304-acre restored prairie, managed by the DNR.

Directions:	From the town of Spring Valley, go 4 miles south on U.S. 63 then 1.5 miles east on 190th Street.
Seasons/Hours:	Year-round.
Length:	N/A
Precautions:	Hunting season.
Amenities:	Hiking, snowshoeing, cross-country skiing.
Information:	Website: www.dnr.state.mn.us/wmas/index.html and search for Goethite.

MASONIC PARK CAVE
FILLMORE COUNTY

Ask around in Spring Valley about the existence of a cave in a Fillmore County park, and you may get blank looks. Many residents who do know of the cave at Masonic Park have never looked for it, being warned off as children by parental alarms of rattlesnakes.

Only five miles northeast of town, Masonic Park lies across Deer Creek from a forested bluff in the Stewartville and the Prosser Members of the Galena Formation. After bordering the approximately 90-foot cliff for roughly 0.1 mile, the creek pulls away from the bluff, which contains Masonic Cave.

To reach the cave, park near the shelter and walk back on the road 0.1 mile, re-crossing the bridge over the

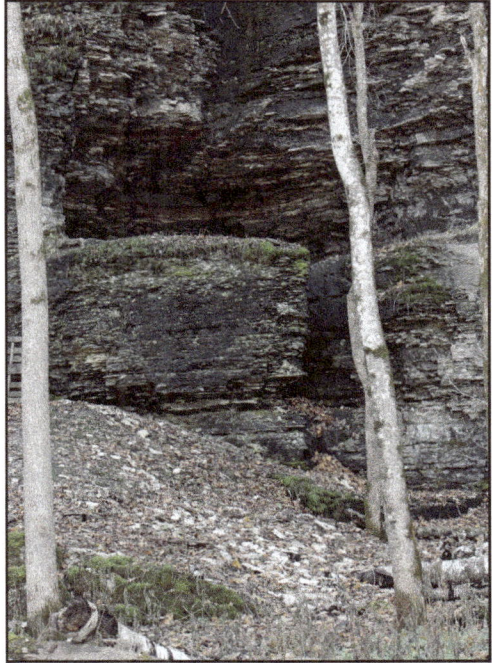

The entrance to Masonic Park Cave is on a ledge about twelve feet above ground. (Photo: Doris Green)

The view inside Masonic Park Cave. (Photo: Shawnee Hoffman)

creek. The cave entrance can be seen in the bluff face during leaf-off conditions. Look for a path on your right near the S-curve sign before the curve. Follow this trail toward the bluff. (Look for jack-in-the-pulpits and bluebells in late May and early June but be wary of nettles in late summer.) When you reach the face of the cliff, look up. The broad cave entrance is perhaps a dozen feet off the ground.

One approach is from the right along narrow "steps" and a three-inch ledge. But often the easiest way to enter this cave is to have a friend boost you up. The entrance room is about 10 square feet. At the rear is a belly-crawl passage best left for hardcore cavers.

Development of the 8.5-acre property dates to the establishment of a sawmill on the site in 1865. Following the death of mill owner George Weisbeck, the parcel was sold and in 1915 was deeded to the Spring Valley Masonic Lodge 58 by one of its members, Leonard Siebert. The Masons built the pavilion and picnic tables, and maintained the property for many decades; however, by the late 20th century their declining membership and a rise in vandalism made keeping up the park difficult. The Masons gave the park to Fillmore County in 1999. No longer visited as often as in the early 20th century, the park is not entirely forgotten, and more than one couple has exchanged marriage vows there at the picturesque bluff-side creek.

Directions:	From Spring Valley, turn onto County Highway 1 and follow it north 1.0 mile to County Highway 38 (gravel). Turn right (east) and follow 38 around several ninety-degree curves (which track section lines) approximately 3.0 miles to the park, on the left.
Seasons/Hours:	Year-round.
Length:	Cave entrance room measures about 10 square feet.
Precautions:	The cave entrance is located in a bluff face about 12 feet above the ground.
Amenities:	Picnic shelter, tables..
Information:	Trip Advisor website: www.tripadvisor.com and search for Masonic Park, Spring Valley, MN.

EAGLE BLUFF ENVIRONMENTAL LEARNING CENTER
FILLMORE COUNTY

The karst exhibit filling the Schroeder Visitor Center at the Eagle Bluff Environmental Learning Center makes southeastern Minnesota's landscape as real and important as the water issuing from your kitchen faucet. Educational panels and interactive displays hammer home the connection between human behaviors above ground with their subterranean impacts. A walk-through replica of an audibly dripping solution cave engages the senses.

The Eagle Bluff Environmental Learning Center connects visitors to their environment, above and below ground.
(Photo: Doris Green)

The exhibit explains the development of Minnesota's nine-county karst country and the implications for the people who live there. Visitors can learn about the impact of problems such as groundwater contamination, landfill leaks, sewage drainage, and waste dumping. They learn one fifth of the United States land surface is considered to be karst and that one in 10 people worldwide live on a karst landscape. The final panels advise against putting anything into the earth you would not want to drink and suggest behaviors to protect groundwater—from testing wells to minimizing the use of fertilizers and advocating for safe land management.

Using its 100 acres and adjacent public lands, Eagle Bluff aims to empower people to care for the earth and offers accredited environmental education programs for individuals, families, and students. Visitors come to hike, ski, and challenge themselves on a high ropes course and in a variety of outdoors skills classes. The center also hosts overnight outdoor learning experiences for students, summer camps, business conferences, reunions, and retreats.

Below ground, the Eagle Bluff property contains a half-mile tunnel through the limestone bedrock. Excavated in 1914, the tunnel directed water from a dam on the Root River through a ridge to the Brightsdale hydroelectric plant that provided power to Preston, Canton, and Harmony. According to the Fillmore County Historical Society, a crew of 300 Bulgarians dynamited the tunnel for the Root River Power and Light Company and lived in bunkhouses at the site.

Until a few years ago, campers in some Eagle Bluff programs were able to explore the tunnel. Today, however, it is a bat sanctuary and closed to the public.

Directions:	From Fountain, follow brown directional signs to turn east on County Road 8 and drive 7 miles. Turn left on County Road 21 and go 1.25 miles. Turn right on Goodview Drive and go 2.25 miles to the Eagle Bluff Campus.
Seasons/Hours:	Visitor center is open weekdays, 8 a.m. to 4:30 p.m., year round and Saturday from 10 a.m. to 5 p.m., June through August. Closed Sunday. Trails are open year round, dawn to dusk.
Length:	N/A
Precautions:	N/A.
Amenities:	Meeting spaces, dining, and lodging; public geocaching course; hiking and cross-country skiing trails; popular high ropes course offered in summer.
Information:	Eagle Bluff Environmental Learning Center, 28097 Goodview Drive, Lanesboro, MN 55949. Phone: 507-467-2437; 888-800-9558—toll Free in MN, WI, IA. Website: http://www.eagle-bluff.org/.

RENO CAVE
HOUSTON COUNTY

Incongruous as a skyscraper, the "palisades" stand on Reno Bluff high above the Mississippi River. A half dozen vertical sandstone struts present a sheer façade in contrast to the cliff's rougher joints and columns. Inside the small palisades, Reno Cave stretches back into the bluff, like La Moille Cave and other sandstone caves found along Minnesota's deep river valleys.

Half the fun of exploring Reno Cave is getting to it. Located in the Reno Management Unit of the Richard J. Dorer Memorial Hardwood Forest, the trail begins at the Reno Horse Campground on State Forest Road 540. This road runs north off Hillside Road, which heads west from State Highway 26 near the town of Reno. The campground is about a mile off the highway.

While the hike from the campground to the cave is only 1.2 miles, it entails a climb of approximately 300 feet. Soon after beginning this hike with a companion, Doris almost placed her hiking boot down atop a mole. On a mid-November day, the mole's velvety gray fur perfectly camouflaged its rotund body among the fallen leaves. Pausing to watch the mole burrow among the leaf litter gave Doris (whew!) a chance to catch her breath.

The trail ends on the roof of Reno Cave, from where three states can be seen. (Photo: Doris Green)

Hannah Lieffring at the base of the Reno Cave palisades. (Photo: Doris Green)

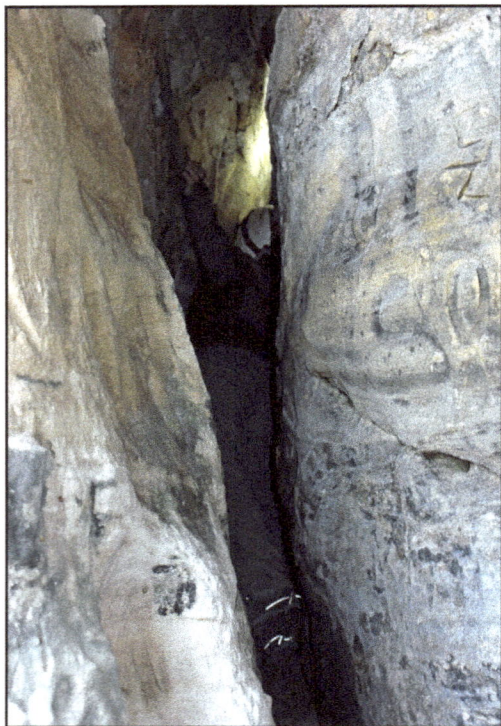

Slithering into the crack at the rear of Reno Cave (Photo: Doris Green)

The forest offers several trails to choose from—keep hiking up and south to reach the cave. On the top of the ridge that runs toward the Mississippi, a campsite (coordinates 43.6016097, -91.2866520) provided a meditative respite to hikers who opt to stay the night.

Our trail abruptly ended at the top of the cliff overlooking the Mighty Mississippi (coordinates 43.5984968, -91.2830270). So where was the cave? Doris wondered. Immediately to the south a rugged rock outcrop faced the river. The cave must be somewhere below us.

Indeed. We were standing, we soon realized, on the limestone roof of Reno Cave. While you might rappel down to the cave entrance, my companion found an easier route by backing up along the trail and then angling down via a deer trail toward the entrance.

Entering the cave proved equally easy. No need for a rope or ladder. Visitors can simply step up into the entrance between two of the palisade pillars.

Inside, a five-by-eight-foot room rises at least 10 feet overhead. Turning around, the entrance offers a view of the broad valley, the river, and the chimneys marking the retired nuclear and active coal power plants at Genoa, Wisconsin. A smaller crevice opening presents a more limited river outlook, and high on the room's north wall, a square window provides a view of seasonally colored woodlands.

Like La Moille Cave in Winona County, Reno Cave reportedly once contained pictographs etched into the soft sandstone walls. The most famous of these was called the Reno face—a human face carved in relief. Today the wall art takes the form of graffiti carved into apparently every square inch of the friable stone. When we visited, a couple of hollowed-out wall pockets sheltered Asian ladybugs and mosquitoes.

At the back of the room, a floor-to-ceiling crack extends farther into the bluff. Thin, fit folks can slither through the crack and climb to an approximately four-by-five foot room with a ceiling height of about five feet. Look for tan camel crickets in the small space, which contains neither graffiti nor pictographs.

Despite the magnitude of the graffiti in the main room, Reno Cave and its environs were largely free from trash and other evidence of partying. The partiers probably tend to stay in the parking area, which offers the convenience of pit toilets and a flat space to build a bonfire.

Directions:	From State Highway 26 at Reno, take Hillside Road west to State Forest Road 540 and drive about a mile to the parking area. Coordinates for the side of the cave are 43.5986427, -91.2829126
Seasons/Hours:	Year-round, dawn to dusk.
Length:	Hike 1.2 miles to the cave, which contains a five-by-eight foot room.
Precautions:	Hiking uphill to the cave requires physical effort and agility. Wear blaze orange/pink during hunting seasons.
Amenities:	Primitive campsite at the top of the ridge.
Information:	The Reno Horse Campground is managed by Beaver Creek Valley State Park, 15954 County 1, Caledonia, MN 55921. Phone: 507-724-2107.

BEAR CAVE PARK
OLMSTED COUNTY

Once upon a time, back in the mid-20th century, the land comprising Bear Cave Park served as a pasture for a dairy herd. Olmsted County acquired the farmland around 1972. The county developed the parcel, constructing roads and picnic shelters, and then named it "Bear Cave Park." Twenty years later, Olmsted County transferred ownership of the park to Stewartville, which continues to maintain it.

Located just north of the Root River, the park offers playing fields and hiking trails, but no obvious bear cave. Past the playing fields, woods skirt the Root River. If you follow

BIOLOGY BELOW

The native, tan camel crickets of Reno Cave are only one of the more unusual life forms that may be encountered in a Minnesota cave. These crickets belong to the Rhaphidophoridae family within the Orthoptera order, alongside grasshoppers and katydids. Camel crickets are up to almost an inch long and hump-backed, with long antennae and legs enabling them to jump several feet—and startle unwary cavers. They lack wings, so are chirpless. Species living in deep, completely dark caves may have reduced or missing eyes. They will eat anything organic, including smaller insects.

Larger insects such as cockroaches may also be seen in caves, and several species of strictly cave spiders have so far been classified in the United States, for example, in 2012 cave explorers discovered a new family of spiders in the Siskiyou Mountains of Oregon. Scientists dubbed it "Trogloraptor" (Latin for "cave robber") because of its formidable front claws. More often, cavers encounter spiders that might be found in basements, under logs, or in other dark spaces. Like camel crickets, they may eat gnats or other small insects attracted by fungus or other organic material, particularly near cave entrances.

Polluted urban caves sometimes contain a sort of fly-and-worm ecosystem. Found deep under Minneapolis, the Schieks Cave biota is a guanophile (excrement-loving) community, with earthworms covering the floor like spaghetti near the broken sewer lines. Fungus gardens, fed upon by swarms of fungus gnats, in turn support the spiders in the cave. It's truly a concrete jungle in there!

Crayfish and fish can be found in cave streams, with some species simply swimming through underground channels like those in Mystery Cave, and others living more specifically in underground streams and ponds. For instance, the blind Ozark cavefish can be found in Arkansas, Missouri, and Oklahoma. Blind cavefish are generally found in regions where bats and other life forms provide organic material and food year round, but have not been found in Minnesota. Generally, glaciated states like Minnesota have an impoverished cave fauna compared to the unglaciated states south of the Ohio River.

Aboveground, cavers hiking in wild areas may come across skunks, lizards, and a variety of snakes, such as the timber rattler, black rat snake, fox snake, blue racer, and more. Aside from the rattlesnake, these do not present any real danger. Several, however, are threatened, endangered, or species of special concern, so leaving them alone is recommended practice.

the asphalt trail into the woods beyond the park, you will come to a section of the trail that runs on the bank above the river. Here, in low water conditions, you might find an eroded indentation below in the limestone bank. Could this groove once have been large enough to shelter a hibernating bear? Who knows—sometimes even fairy tales carry a grain of truth. And water, wind, and freezing and thawing cycles reshape many riverbanks over time.

Directions:	From Rochester, follow U.S. 63 south to Stewartville. Turn right on County Highway 35 and watch for the park on your left. Bear Cave Park is located in at 800 10th St. NW, Stewartville, MN 55976.
Seasons/Hours:	Year-round.
Length:	N/A
Precautions:	Signs of riverbank erosion may be more visible in late summer and at other times of low water.
Amenities:	Softball fields, tennis courts, disk golf, hiking, picnicking.
Information:	Stewartville Parks and Recreation Department, 105 East 1st St., Stewartville, MN 55976. Phone: 507-533-4745. Website: http:// http://stewartvillemn.com. Hold your cursor over the Community tab to see a list of links to city parks.

QUARRY HILL PARK AND NATURE CENTER
OLMSTED COUNTY

Vestiges of another way of life linger along the trails and half-hidden in the woods of Quarry Hill Park and Nature Center in Rochester. The city purchased most of this 270-acre parcel in 1965 from the state, which had operated the Rochester State Hospital here since the 1870s. Hospital workers and residents—who included the indigent, chemically dependent, and mentally ill—farmed 1,000 acres, growing food for themselves and their livestock.

In 1882, Thomas Coyne led other residents in hollowing out a horseshoe-shaped cave in St. Peter Sandstone as a food storage area. Heavy wooden doors helped stabilize the inside temperature. The residents cut about 20 bins along the sides of the 200-foot tunnel, as well as two larger rooms for storing apples and butter. One year, the bins held 9,300 bushels of potatoes, 417 bushels of rutabagas, 400 bushels each of carrots and beets, and more than seventeen tons of cabbages. The U-shaped tunnel was more than eight feet wide and just as tall—large enough for workers to drive a wagon inside to unload the produce. Later, residents added a 100-foot extension, creating a larger, Y-shaped storage area.

Main gate into Quarry Hill Cave.
(Photo: Doris Green)

Even after canning replaced cold storage preservation, the Hospital continued to use the cave. For example, in 1941, workers stored 20 barrels of dill pickles and 15 tons of sauerkraut in the cave.

After the underground storage area was no longer used, the doors were removed and erosion began to take a toll on the cave. In 1992, two limestone block walls and a gate were installed at the three entrances in order to preserve the cave. Small openings in the block walls provide access for insects and hibernating bats. The steel gate is opened for human visitors during regularly scheduled tours and classes.

Held twice a month, the cave tours generally end with an optional hike to another historic feature in the park, for example, to a pond, reservoir, or the rock crusher retaining wall. This restored wall below the older of two quarry sites is now used as a rock climbing wall.

Workers and residents produced crushed rock from the quarried stone for the Hospital and local contractors, who used it in road and building construction. The stone was transported to the crusher site in narrow gauge railroad cars pulled by horses. The Hospital operated two crushers and a pulverizer, which provided lime dust that farmers spread on their fields to neutralize acidic soil.

Sometimes the cave tours conclude with a hike to a small mortuary cave, which Coyne dug on the east side of a cemetery containing more than 2,000 graves. When hospital residents died during the winter, the bodies were stored in the cave until the ground thawed and graves could be dug in the spring. Now debris-filled and inaccessible, this is the only known, publicly visible mortuary cave in Minnesota.

Quarry Hill Cave is a gigantic root cellar dug into St. Peter Sandstone. (Photo: Quarry Hill Nature Center)

One of two sealed bat entrances into Quarry Hill Cave. (Photo: Doris Green)

Northeast of the mortuary cave, you can find the newer of the Hospital's two limestone quarries, which was worked from the 1890s to 1950. One trail leads around the rim of this quarry, where you can see cracks in the rock face and occasionally the fossils of cephalopods.

An even newer "underground" attraction exists in the nature center, right along with exhibits on geology and wildlife—like a live snapping turtle, catfish, tree frog, and seven of Minnesota's seventeen native species of snakes. The replica solution cave features "rocks" for kids to stand on to capture the perfect photo and audible dripping water.

Directions:	Located on Silver Creek Road, off of County Highway 22 (East Circle Drive).
Seasons/Hours:	Monday through Friday, June 1 through August 31, 8:30 a.m. to 4:30 p.m. and 9 a.m. to 5 p.m. the rest of the year; Saturday: 9 a.m. to 5 p.m.; Sunday, 12 to 5 p.m.
Length:	The hike from the nature center to the storage cave takes about 10 minutes, but most visitors want to allow at least several hours to fully explore this historic and natural site, along with the nature center exhibits.
Precautions:	Portions of the hike to the food storage cave are hilly and not wheelchair- or stroller-friendly.
Amenities:	Environmental classes, picnic shelter, playground, nature center, library, fishing, biking, baseball diamond, tennis courts, cross-country ski and snowshoe rental.
Information:	Quarry Hill Park, 701 Silver Creek Rd., NE, Rochester, MN 55906. Phone: 507-328-3950. Website: https://qhnc.org.

WHITEWATER STATE PARK
WINONA COUNTY

Geology is the watchword at Whitewater State Park. From the park's trails and streams you can see many rock layers along its steep bluff faces. The oldest exposed rock here dates to more than 500 million years ago—not so very long ago to geologists, who date the earth at about 4.6 billion years.

About 450 million years ago, most of what today is southeastern Minnesota, southwestern Wisconsin, and Iowa lay beneath a shallow sea. Sediments built up on the sea floor and were cemented together over eons to form rock hundreds of feet thick. When sand washed into this sea, layers of sandstone were formed. When the sediment contained the remains of many ancient sea animals, limestone was the result. When the sediment contained many particles of clay, shale was created.

When the sea withdrew, water and wind eroded hills and valleys. During the more recent Ice Age, which ended a scant 10,000 years ago, glacial meltwater further eroded

BANSHEE CAVE SUBTERRANEAN LABORATORY

Greg was exploring the Ice Box, a peculiar cave on the banks of the Root River near the Fugles Mill Historical Site, on August 6, 1989, when he stumbled upon another, even larger cave. The unsettling appearance of the passages, which appeared ready to collapse with an ominous roar, led him to name it Banshee Cave. It's a superb example of a "mechanical cave," formed by gravitational slumping, rather than by groundwater dissolving out the passages. The estimated 50 yards of passages form a rectilinear maze, much of it described on his sketch map as "muddy belly crawl," patroled by raccoons, whose glowing eyes could be seen in the distance.

One of the passages in Banshee Cave was of walking size and much drier. Greg decided that this would be the ideal location for a proposed "subterranean laboratory" where the cave life of the region could be studied in its natural habitat. According to his manuscript notes, "I was inspired to undertake this project after reading a chapter on subterranean laboratories in Vandel's *Biospeleology*." (Albert Vandel wrote this famous French textbook about cave life, published in 1965.) On August 19, 1990, Greg transported a load of supplies through the tight crawlways, leaving a table and chair for the convenience of researchers. Perhaps the most notable observations made involved the minute insects known as springtails. While activity soon ceased owing to the danger and difficulty of routine access, the table and chair remain there today so far as Greg is aware! A unique sort of laboratory without parallels in Minnesota.

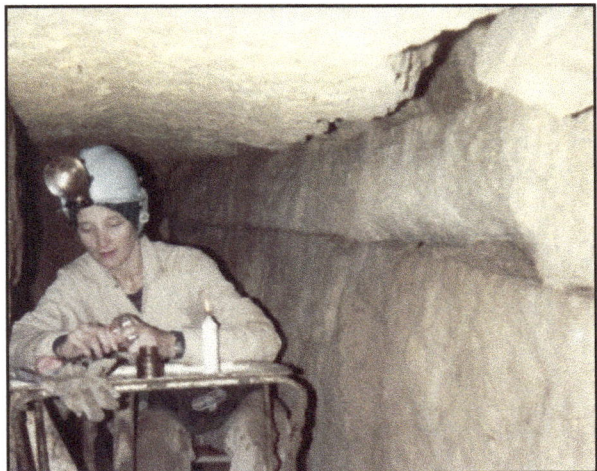

Banshee Cave was equipped with chairs and tables for the study of cave life. (Photo: Greg Brick)

away the top layers of soil and rock and carved into the older layers, leaving the bluffs you see in the park today. If you walk to the top of Coyote Point, you will be standing on the Shakopee Formation. Named for Shakopee, Minnesota, where this layer is well exposed, the Shakopee is mostly dolostone (magnesium-rich limestone) plus sand and shale.

Beneath the Shakopee, you can see Oneota dolostone as you hike up and climb down from the Coyote Point summit. Oneota is mostly dolostone, with less sand and shale than the Shakopee. This layer often contains fossils and caves, like **Coyote Point Cave**, clearly visible (though inaccessible) from the trail to the point. Nearby, several smaller cavities appear in the bluff face.

Across the Whitewater River valley, another Oneota pinnacle points skyward. You can climb to this pinnacle, Chimney Rock, and crawl into the eroded cave at its base. **Chimney Rock Cave** features four windows—two overlooking the valley—and is large enough for an adult to sit up. (Children enjoy the fort-like appearance of the cave, but the sharp drop-offs beneath the two windows make adult supervision a must.) From Chimney Rock Cave, you can see Coyote Point, and some geologists believe that long before the glacial meltwater cut through this valley, Chimney Rock Cave and Coyote Point Cave were once both part of the same large cave.

If you hike down to the Whitewater River from Chimney Rock, you will see Jordan Sandstone along the bank. The rust-colored sandstone bears the carvings of early tourists, though, understandably, defacing the natural rock face is not permitted today.

Whitewater State Park contains one other significant cave. **Trout Run Creek Cave** is inaccessible; it requires rock-climbing equipment to reach and rock climbing is not permitted in the park. Still, there are other intriguing geologic features to search for in addition to Chimney Rock Cave—including smaller cavities, a few springs, and the spot where a stream disappears into the earth.

One hundred years old in 2019, the park preserves human history as well as the rocks and fossils documenting geologic time. It contains 29 structures built by the Civilian Conservation Corps (CCC) and Works Progress Administration (WPA) in the 1930s. The timber and limestone structures range from the entrance sign to a shelter to a footbridge and dam. Much of the limestone used in construction was quarried within the park.

Two miles east of the park, the Elba Fire Tower was completed by the CCC in 1936. Located just east of the town of Elba, the 110-foot tower is an interpretive site open during daylight hours from April through October. You need to climb more than 500 steps just to reach the base of the tower, which has a seven-by-seven-foot observation enclosure at the top.

Directions:	From Rochester follow U.S. 14 east (or from Winona follow U.S. 14 west) to State Highway 74. Turn north on Highway 74 and drive 18 miles to the park.
Seasons/Hours:	Year-round, 8 a.m. to 10 p.m., however, water is available only from a pump near the visitor center from late fall through early spring. The park office is generally open from 9 a.m. to 4 p.m., with longer hours in the summer.
Length:	Trail lengths vary. The Chimney Rock Trail is less than 0.7 mile in length and of moderate difficulty, with some steps.
Precautions:	The Coyote Point Trail is challenging, involving switchbacks and a climb to the bluff top.
Amenities:	Whitewater State Park offers a year-round visitor center, gift shop, rustic cabins, campgrounds, trout fishing, swimming beach, picnicking, kayaking, hiking, cross-country skiing, and snowshoeing. Winterized cabins can be reserved. Geocaching is also popular and you can borrow GPS units at the park office.
Information:	Whitewater State Park, 19041 Highway 74, Altura, MN 55910. Phone: 507-312-2300. Website: www.dnr.state.mn.us/state_parks/index.html and use the Park Finder.

THE GAINEY GOLD MINE FRAUD

On the outskirts of Whitewater State Park, cattle make use of the natural air conditioning provided by an abandoned mine entrance in a farmer's field. In 1910, according to Roger Kehret's 1974 booklet, *Minnesota Caves History and Legend*, a gold mine was opened at this location and a company was incorporated, selling shares around the United States. In 1925 the fraud was exposed, after the perpetrators had made a fortune from the swindle. While small amounts of genuine gold have been found in southeastern Minnesota, it's never been enough for a sustained mining operation.

DEVIL'S CAVE, BLUFFSIDE PARK
WINONA COUNTY

Despite having such a diabolical name, getting to Devil's Cave requires a long trek uphill into the heavens. High up a ravine in the colorful sandstone bluffs above Winona is a long, narrow, natural cave in a city park, featured on early postcards. Developed along a vertical joint in the Jordan Sandstone, 80 feet long, it requires gymnastic maneuvers to get to the so-called party room at the far end of the cave. How the cave got its name is a mystery, but Greg once found an amulet back there, and from the accumulation of tell-tale detritus he could tell that it's a party cave for the locals, if that be devilish. The Holzinger Trail, starting at Holzinger Lodge, on West Lake Boulevard, winds through Bluffside Park and offers some nice views of the city, but it should not be confused with the trail leading to the cave, which lies to the west.

Directions:	Bluffside Park, 925 West Lake Boulevard, park at the pullout. UTM coordinates for the cave are 606682 E, 4877194 N.
Seasons/Hours:	Year-round, dawn to dusk.
Length:	One quarter mile hike uphill from the roadside pullout to get to the cave, which is 80 feet long.
Precautions:	The quarter-mile hike uphill to the cave involves 340 feet of elevation gain and traversing the cave requires physical effort and agility.
Amenities:	The nearby Holzinger Trail System.
Information:	Winona Parks, 105 City Hall, 207 Lafayette St, Winona, MN 55987

Devil's Cave developed along a joint in the Jordan Sandstone. (Photo: Greg Brick)

Devil's Cave in 1909. (Photo from the Greg Brick Collection)

TREASURES UNDER SUGAR LOAF
WINONA COUNTY

Hidden behind a red velvet curtain, an historic former brewery cave adds to the allure of a Winona antiques store, Treasures Under Sugar Loaf. The store occupies the former Bub's Brewery at the base of Sugar Loaf, a limestone outcrop reshaped by quarrying. Despite the insulating properties of a plywood panel installed behind the curtain, the manmade cave helps cool the basement of the antique store in the summer and warm it in the winter.

Really a mall, the antique store offers an amazing array of small collectibles, crafts, and home décor from more than 60 dealers. Each item has its own story to tell but none is more noteworthy than the story of the cave and the building above it.

Founded in 1856 by Jacob Weisbrod, brewery operations were moved to the base of Sugar Loaf in 1862. After Weisbrod died of typhoid fever in

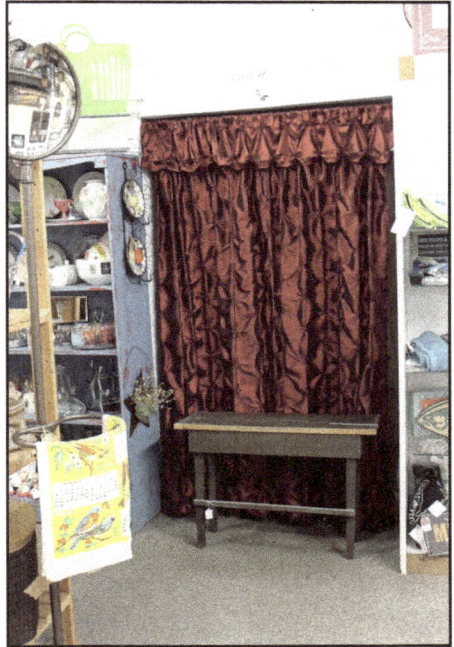

A red velvet curtain hides the closed entrance to the cave behind the Treasures Under Sugar Loaf antiques mall. (Photo: Doris Green)

1879, Bavarian immigrant Peter Bub took over. He married Weisbrod's widow and expanded the plant, including digging the cave, which includes several rooms. Bub (pronounced "Bube" or "Boop" as in "Betty Boop," depending on whom you ask) ran the business until his death in 1911. The firm stayed in the family, with the next owner, William Miller, who had married Bub's daughter, Lena, in 1899.

The brewery prospered, shipping its products across southeastern Minnesota and as far as the Twin Cities, Iowa, and Wisconsin. In contrast to several other Winona breweries, Bub's survived the Great Depression by producing soda and "near beer," with a low alcohol content.

Still, the growth of large national breweries and transcontinental trucking led to increased consolidation in the brewing industry, and Bub's finally ceased production in 1969. Nine years later it was listed on the National Register of Historic Places for its local significance in the industry.

No longer publicly accessible, the Bub's Brewery cave nevertheless still exists behind that red curtain. When stocked with blocks of lake ice, the cave passages, with their limestone walls, offered premium beer storage. Slanted floors made cleaning up spills easy. A shaft, possibly for ventilation, opens in a passage ceiling and heads toward the bluff top.

Bub's name still exists, too, in the form of Bub's Brewing Company, 65 East 4th St., a bar/restaurant in downtown Winona. This Bub's, which opened in 1992, proudly serves beers like Bub's Black Forest Ale, Bub's Amber Red, and Bub's Golden to keep the name and history alive for both locals and visitors.

Directions:	On the frontage road southwest of U.S. Hwys. 14 and 61 in Winona.
Seasons/Hours:	Open from 10 a.m. to 5 p.m., Thursday through Monday, plus on Tuesday, June through August.
Length:	N/A
Precautions:	N/A
Amenities:	Winona, the "island city," offers a diverse selection of restaurants, as well as lodging, educational facilities, and recreational opportunities, such as trails up Sugar Loaf.
Information:	Treasures Under Sugar Loaf, 1023 Sugar Loaf Rd., Winona, MN 55987. Phone: 507-474-7030. Website: https://www.treasuresundersugarloaf.com/

LA MOILLE CAVE EXHIBIT, WINONA COUNTY MUSEUM
WINONA COUNTY

As much about history and archaeology as geology, the La Moille exhibit at the Winona County Museum replicates an underground site discovered in Jordan Sandstone. Inside the replica, petroglyphs recorded in 1888 by Theodore Lewis are recreated on the walls, including fish, rattlesnakes, raccoons, human figures, and thunderbirds.

Lewis discovered the rock art in **La Moille Cave**, named for the nearby town of La Moille (rhymes with "oil"). Evidence suggested that the cave served as a Native American hunting camp during the Woodland period, approximately 800 B.C. to 900 A.D.

Confusingly, in addition to the Woodland period La Moille Cave, there existed a **La Moille Rock Shelter** dating to about 1500 B.C. Located about a quarter mile southeast of La Moille Cave, the rock shelter served as a fishing camp. Discovered—and then destroyed—by road workers on U.S. Highway 61 in the 1950s, it contained projectile points, bone fragments, and "some of the earliest pottery known in Minnesota," as Greg reported in his book, *Minnesota Caves: History and Lore*.

The La Moille underground replica at the Winona County Museum features several pictographs, including this thunderbird. (Photo: Doris Green)

Though much impacted by the construction of the nearby Lock & Dam No. 6 on the Mississippi River in the 1930s, La Moille Cave still exists. Now a shadow of its former self and nearly filled with sediment, the cave has only a little clearance and does not present a welcoming environment. "A spring at the rear of the cave guarantees a mud bath to any visitor," Greg noted. The petroglyphs have weathered away.

Not surprisingly, the two similarly named sites so close together sometimes confused later explorers and archaeologists. In fact, when construction on U.S. 61 obliterated the rock shelter site, some believed the road construction had destroyed the cave, as the shelter became confounded with the cave.

Directions:	From U.S. 61, turn onto Huff Street at Lake Winona. Turn right onto 4th Street and drive to Johnson Street; parking will be to your right and the History Center will be on the left.
Seasons/Hours:	Open weekdays, 9 a.m. to 5 p.m., Saturday 10 a.m. to 4 p.m. and Sunday, 12 to 4 p.m. Closed Holidays.
Length:	N/A
Precautions:	N/A
Amenities:	Winona, the "island city," offers a diverse selection of restaurants, as well as lodging, educational facilities, and recreational opportunities, such as trails up Sugar Loaf.
Information:	Winona County Historical Society's Winona County History Center, 160 Johnson St., Winona, MN 55987-3434. Phone: 507-454.2723 ext. 0. Website: www.winonahistory.org/museums.html.

KRUEGER'S CAVE
WABASHA COUNTY

Krueger's Cave is not for claustrophobes. Its maze of tight, dusty crawlways can be confusing and Greg even got lost in the cave years ago. It was this aspect that led Boy Scouts to leave string through the thousands of feet of labyrinth, remnants of which can occasionally still be seen. At the far end of the maze is the Calypso Room, a welcome relief, full of stalactites and stalagmites, setting it apart from the rest of the cave, which is pretty barren.

Located on a finger ridge of private land along West Indian Creek, near the town of Plainview, Krueger's Cave is accessible by contacting the Minnesota Speleological Survey, which, like the Minnesota Caving Club, is a grotto (member organization) of the National Speleological Society.

The Minnesota Speleological Survey meets monthly in Bloomington and hosts two annual events: a Sleepunder in July and a Cornfeed in September. Membership includes geologists, engineers, and other professionals, as well as people whose hobby is caving. Members—all are volunteers—may get involved in cave exploration and surveying; karst, and groundwater protection; cave rescue training; biological studies; reconnaissance of caves and sinkholes, at the request of landowners; and other aspects of karst studies.

Directions:	N/A
Seasons/Hours:	Hours: Tours scheduled through the Minnesota Speleological Survey.
Length:	The cave is about 0.67 miles long.
Precautions:	Helmet, light sources, boots, coveralls or other rugged clothing recommended.
Amenities:	N/A
Information:	Minnesota Speleological Survey. Website: https://sites.google.com/view/msscaves.

MAZEPPA CAVE
WABASHA COUNTY

Native Americans said there was a spirit lurking in this colorful natural cave in the Jordan Sandstone on the outskirts of the small town of Mazeppa, south of the Twin Cities. Settlers took this to mean a devil. For many years, the 70-foot-long Mazeppa Cave has been used as a root cellar, sealed behind a door by the landowner, although a team of archeologists investigated this site in 1997. They found cave walls grooved by many petroglyphs. A narrow tunnel at the back of the cave might be a symbolic entry point into the earth, where a shaman could consult with the "rock people" about which mineral remedy to use to treat his patients. More so than any other petroglyph cave in Minnesota, Mazeppa would appear associated with shamanistic practice.

Directions:	From the town of Mazeppa, head east on State Highway 60, then south on County Road 7, then west on 603rd Street for 0.5 mile, to see the door in the sandstone outcrops on the north side of the road.
Seasons/Hours:	Year-round.
Length:	70 feet.
Precautions:	Cave is on private property; do not disturb landowner.
Amenities:	N/A
Information:	N/A

Mazeppa Cave in the colorful Jordan Sandstone has some colorful stories to tell. (Photo: Greg Brick)

FRENCH SALTPETER CAVES, FRONTENAC STATE PARK
GOODHUE COUNTY

The extended French presence in the Upper Mississippi valley required gunpowder and the usual assumption (valid in most cases) is that it was imported from France. A vital constituent of gunpowder is saltpeter (potassium nitrate), a mineral sometimes obtained from caves. The French fur trader Pierre-Charles Le Sueur (1657–1704), while ascending the Mississippi River in 1700, reported saltpeter caves in his journal, which remained unpublished for many years. This is the earliest reference to caves in what is now Minnesota.

Le Sueur's comments about the caves being inhabited by bears in winter and rattlesnakes in summer suggests that they were visited (by someone) throughout the year, and presumably there would have been a good reason for this. Although Le Sueur described the Lake Pepin caves as containing "saltpeter," he was more likely referring to a precursor substance (calcium nitrate), not potassium nitrate. The prevailing humidity in Minnesota caves is too high for him to have encountered anything other than deliquescent salts (dissolved in the sediment) rather than the crystallized saltpeter seen in desert

Vista of Lake Pepin as seen from a saltpeter cave in Rattlesnake Bluff. (Photo: Greg Brick)

regions. Apart from whitish snow-like efflorescences, not even experienced saltpeter prospectors could identify nitrate-rich sediments by sight and the usual confirmation was a bitter taste and other subtle clues until modern chemical tests for nitrates were developed. But this precursor substance could be easily converted to saltpeter.

The location of Le Sueur's saltpeter caves, near Red Wing, as inferred from his journal, was examined more than 300 years later, in 2004. Sediment samples collected there and from other rock crevices in the Lake Pepin bluffs revealed a high nitrate content (up to 3.5 percent) in the laboratory. This compares favorably with the nitrate content of sediments from known American saltpeter caves in Kentucky, which range between 0.01 percent and 4.0 percent nitrate. Even though no mining tools or indications of saltpeter mining were observed in any of the Minnesota caves, the abundant nitrate (likely of organic origin) in cave sediments at the approximate location described by Le Sueur corroborates that there was a kernel of truth to what he reported in his journal.

No artifacts from the fur trading era have been found in the small limestone caves at Frontenac State Park, yet they do contain sediments with high nitrate concentrations, which had the potential to be used in gunpowder making. Upon entering the largest such cave in Rattlesnake Bluff, which is barely large enough to turn around in, you'll notice the dry reddish soil, usually dappled with raccoon footprints and the occasional buzzard feather. Often the walls are found encrusted with cinnamon (brown) quartz. A playful aspect of caves of the right shape is that they are like cupping a seashell to your ear. Except in this case you're inside the seashell! Greg happened to notice this one windy day when the waves of Lake Pepin, far below, were crashing against the shoreline, creating a faint reverberation inside the cave.

Usually these caves taper down to rock tubes large enough only for a raccoon. No one has reported a rattlesnake from this bluff for many years, and after traversing the crevices for several years during saltpeter research, none were seen. So you're safe on that point. And keep in mind that rattlesnakes are a protected species so you're not allowed to remove or kill them anyway.

Cave saltpeter played an important role in American wars, from the Revolutionary War and War of 1812 to the Civil War, where it became a strategic mineral for the Confederate armies, cut off as they were by a Union naval blockade. As late as World War I, natural nitrate deposits were eagerly sought out, but were finally rendered irrelevant by the so called Haber-Bosch process, which manufactures nitrate from the nitrogen of the atmosphere.

Directions:	Rattlesnake Bluff forms the isolated, western half of Frontenac State Park but there's a public parking area at the east end of Lakeview Avenue at Greene Point. From there, hike west back up the road, to where the Rattlesnake Bluff Trail begins, and from that point, instead of hiking the trail itself, ascend directly up the steep talus slope through the trackless woods to the rock outcrop. UTM coordinates for the cave are 547636 E, 4932274 N.
Seasons/Hours:	Year-round, 8 a.m. to 10 p.m., but the caves are more easily found in winter, without vegetation hiding the entrances. Avoid fall hunting season, as this park often has a special deer hunt. Park fees.
Length:	Several feet
Precautions:	These caves are more difficult to get to than any others described in this book, because of the terrain and the effort required. Exercise extreme caution when ascending the steep talus slope as boulders may shift unexpectedly, causing a fall or sprained ankle. Once you have reached the top of the slope, the rock outcrops will have to be negotiated if you're to actually enter the caves.
Amenities:	Frontenac State Park has scenic views of the Mississippi River and is a well-known birding spot. Also check out the historic Garrard limestone quarry, along the park trails.
Information:	Frontenac State Park 29223 County 28 Boulevard, Frontenac, MN 55026. Phone: 651-345-3401. Website: www.dnr.state.mn.us/state_parks/index.html and use the Park Finder.

CARLSON LIME KILN
GOODHUE COUNTY

When you walk up the low hill toward the G. A. Carlson Lime Kiln on Barn Bluff your perspective from below makes the blocky edifice loom even larger than it is. Fifty-four-feet high and 60 feet wide at the base, the immense kiln testifies to the importance of the lime industry in Red Wing, where 30 kilns operated from about 1860 to 1908. Constructed of limestone blocks, the kiln protrudes 60 feet from the bluff and has three arched openings in front. The central opening is 20 feet in height and two side openings, which served as brick-edged fireboxes, are five feet in height. A chimney at the back rises up into the bluff. Limestone was quarried on top of the bluff and the kiln filled through tunnels from above.

A historic marker at the end of East Fifth Street in Red Wing records that Gustavus Adolphus Carlson, who built this kiln in 1882, simultaneously operated six kilns and three quarries in this city. Bluff-top quarries provided both limestone and lime for mortar and plaster. Large wood-fired kilns roasted the limestone at temperatures up to 2,000 degrees Fahrenheit for several hours, driving off carbon dioxide and leaving lime (calcium oxide). Smaller kilns might burn for several days or weeks.

A spur of the Chicago, Milwaukee and St. Paul Railroad fronted the kiln, transporting the lime to distant markets. Workers loaded the lime onto the cars by wheelbarrow.

Constructed in 1882, Carlson Lime Kiln measures 54 feet high and 60 feet wide at the base.
(Photo: Doris Green)

Just east of the tracks, the Mississippi River flows toward Dubuque, Iowa; St. Louis, and New Orleans.

Initially workers quarried the limestone by hand with sledge hammers, chisels, and pry bars. Later they used dynamite, with the larger explosions felt as far as a dozen miles from the bluff. The reverberations led to citizen protests and efforts to preserve the bluff; it was deeded to the city of Red Wing in 1910. The kiln was placed on the National Register of Historic Places in the bicentennial year of 1976. Barn Bluff itself was placed on the Register in 1990.

Near the dead end of East Fifth Street, a concrete stairway heads toward the top of Barn Bluff. Below the crest, a wooden shelter, signage, and benches offer a respite after the climb. A marker explains that early French explorers named the bluff "Mt. La Grange" because it resembled a large barn. Its Native American name, He Mni Can, referred to the hill that appeared to stand in the water.

Barn Bluff is 3,100 feet in length, 800 feet wide, while its highest point is 343 feet above the Mississippi. It formed as an island when the river, then five miles wide, carried torrents of glacial meltwater. A vertical fault exists in the south slope, where a section has dropped 150 feet lower than the remainder of the bluff. The National Register of Historic Places registration form notes the fault has exposed greenish Franconia Sandstone, St.

Lawrence dolomite and shale, and Jordan Sandstone. Trilobites and other marine fossils are also visible.

Farther along the trail, you can see kiln tunnel openings, used to dump wood into the fireboxes. Also above the kiln is a stone-arch tunnel that once extended 30 or 40 feet into the side of the bluff; however, the rear 15 feet or so caved in years ago. This tunnel was apparently used for dynamite storage.

The view from the top of Barn Bluff is well worth the climb. Long before quarrying took over the bluff top, Native Americans and early settlers used it as a lookout. From a distance, as its Native American name noted, the bluff appears to rise from the middle of the Mississippi; its top still offers broad vistas up and down the river.

Early travelers including Jonathan Carver, Zebulon Pike, Henry Schoolcraft, and Henry David Thoreau climbed the bluff and remarked upon the beauty of the valley below. As reported in the National Register of Historic Places registration form, Major Stephen H. Long wrote in 1817, "from the summit of the Grange, the view of the surrounding scenery is surpassed perhaps by very few, if any, of a similar character that the Country, and probably the world can afford."

Directions:	From downtown Red Wing, follow East Fifth Street to the east side of Barn Bluff. From the dead end of the road, the 550-foot Carlson Lime Kiln Trail takes you to the base of the kiln. Walk back toward the parking area and the stairway that leads to the top of Barn Bluff.
Seasons/Hours:	Year-round, weather permitting.
Length:	A hike to the summit and back takes an hour or two, depending on your pace.
Precautions:	Steep stairway and moderately challenging trail.
Amenities:	The city of Red Wing is known for its pottery, antiques, and parks.
Information:	Red Wing Parks and Recreation, 315 West 4th St., Red Wing, MN 55066. Website: www.red-wing.org and use the Community tab to get to the Barn Bluff web page.

OVERLOOK CAVE, RED WING MEMORIAL PARK
GOODHUE COUNTY

The narrow snake of Skyline Drive passes two historic quarries on its way to the top of Sorin's Bluff, suddenly leaving the woods to arrive at a broad prairie, covered with grasses and wildflowers. Look for the round Martello tower, reminiscent of the one at Fort Snelling State Park, as the road circles the plateau to an observation area. A low concrete-topped stone wall keeps visitors from the edge of the cliff. This prominence offers a bird's eye view of downtown Red Wing's streets and buildings, as well as the Mississippi River's network of side channels and sloughs.

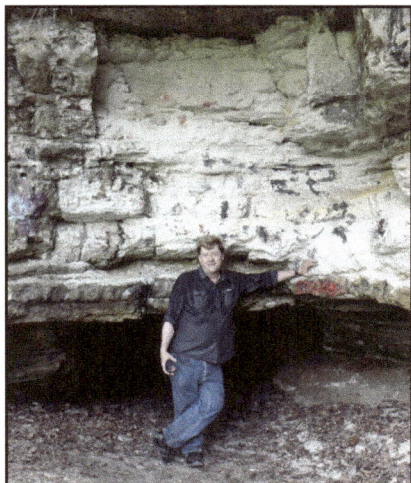

Greg Brick at the entrance of Overlook Cave.
(Photo: Doris Green)

Greg Brick at the back of Overlook Cave. (Photo: Doris Green)

One of two entrances to Horseshoe Cave. (Photo: Greg Brick)

Like Barn Bluff, Sorin's Bluff survived as an island in the large glacial meltwater river because its hard overlying Oneota dolomite prevented erosion. A bronze marker erected by the Geological Society of Minnesota and the city of Red Wing in 1955 describes the impact on the landscape of four major epochs of glaciation. The meltwater "eroded the broad valley of the Mississippi River 200 feet deeper than the present channel." When the ice receded, the volume and force of the water lessened. The river was unable to remove all of the sediment carried in "by its high-gradient tributaries," according to the marker. "Thus the valley was filled to its present level and exhibits a remarkable series of meanders, oxbow lakes, side channels, sloughs, swamps, and tillable land."

Seventy feet beneath the observation area, a manmade cave, called **Overlook Cave**, extends approximately 180 feet into in the Jordan Sandstone. To reach it, follow the rough trail to the right of the stone wall. Half buried pipe railings suggest heavier use of this trail at one time.

Marked by graffiti, the cave's low, wide opening is impossible to miss. When you enter, note the sand floor and dolomite roof. Take a light toward the rear of the cave and look for a ledge, marked by carvings and more graffiti, in the alcove above. Beyond this alcove, the cave requires stoop walking until, at the very back, you emerge with relief into a colorfully stratified dome a dozen feet high. The cave's most interesting natural features are stone ripple marks exposed on the ceiling during excavation, recording a petrified shoreline from the Cambrian Period (485 to 540 million years ago).

Overlook Cave serves as a reminder that this mound was a site of silica mining, as well as limestone quarrying. The trail continues beyond the cave, passing several other small dug caves before turning treacherous and climbing up to the left side of the stone wall at the observation area.

Horseshoe Cave, the other big cave in Memorial Park, can be just as tricky to find. An artificial sandstone cave 200 feet long, its two entrances loop back to each other, hence its name. From the Lower Quarry parking lot, hike the footpath southeastward to where another footpath branches off in a downhill direction, and follow that down to the cave.

Memorial Park's picnic areas feature additional observation points with amazing views of the Mississippi and Lake Pepin. Created in two of Gustavus Adolphus Carlson's abandoned quarries (see BARN BLUFF), the Upper and Lower Quarry Picnic Areas feature a small sugar loaf mound remaining from the quarry days. Established in 1929, the park was upgraded in 2013 with stone picnic tables, informational signage, and fire rings.

Carlson was not the only entrepreneur to engage in the lime industry on Sorin's Bluff. Swedish immigrant Johannes Johnson, for instance, also had a quarry atop Sorin's Bluff and used its stone to build a large house at the bluff's base. Sorin's Bluff also features the remains of several kilns and other quarrying remnants, which you can find along the bluff's recreational trails.

Directions:	In Red Wing, from Seventh Street East, follow the signs to Memorial Park. Turn on Skyline Drive into the park and proceed up the narrow, winding road to the bluff top.
Seasons/Hours:	Spring through late fall, dawn to dusk.
Length:	Overlook Cave is approximately 180 feet long.
Precautions:	The trail to this manmade cave is difficult, particularly in wet conditions. This park does not offer water or electricity.
Amenities:	The bluff top features a nine-hole disc golf course. The historic quarry picnic areas offer access to hiking and biking trails, stone picnic tables, charcoal grills, fire rings, toilets; the Upper Quarry area contains a shelter.
Information:	Red Wing Parks and Recreation, 315 West 4th St., Red Wing, MN 55066. Website: Website: www.red-wing.org and use the Community tab to get to the Memorial Park web page.

JORDAN CREEK
GOODHUE COUNTY

One of the stories from the earliest days of Red Wing involves the biblically named Jordan Creek, which bisected the town. An historical plaque at Jordan Court, bolted to the sidewall of the building at 425 West Third Street, states that "Jordan Creek continues to flow under the streets and buildings of downtown Red Wing. Its course begins beyond Fifth and Plum, carries through the city hall basement area, crosses beneath Jordan Court and enters the Mississippi at Levee Park through storm sewers and general seepage." The marker includes a nice map of the creek. Indeed, you can readily look down to see and hear the noisy creek through manhole gratings in the parking lot behind 418 West Fourth Street, kitty-corner from the marker itself, about 250 feet away. But while you can't actually visit any portion of the stream, ponder the following vicarious narrative of those who did try to follow the underground path of Jordan Creek, starting at the downstream end, where it would meet the Mississippi River.

Back in April 1991, the Goodhue County Historical News carried a brief article titled, "The Catacombs of Red Wing's Sewer System." The anonymous author had explored storm drains leading to Levee Park along the Mississippi River. Reading this, Greg was inspired to visit the stormwater "catacombs" in 2007 along with a long-time Red Wing employee. The hope was to find the elusive Jordan Creek. Most of the passages, they found, were hand-carved in the Franconia Sandstone, which has a greenish tint owing to its glauconite minerals.

The storm drains, though straight overall, meander about somewhat, giving them the appearance of snaking cave passages. The cave-like experience was augmented by the

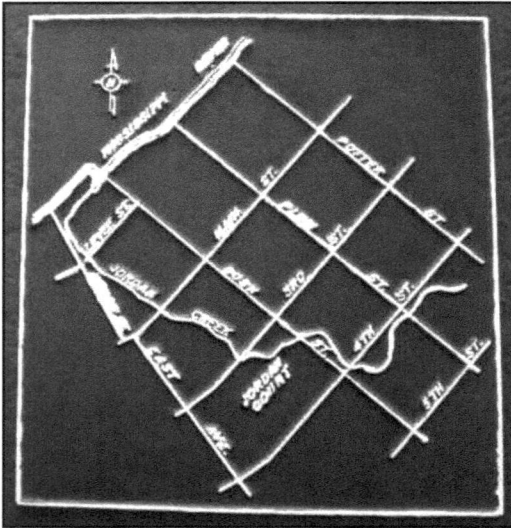

Bronze map showing the subterranean course of Jordan Creek through downtown Red Wing. (Photo: Greg Brick)

mineral deposits, as some parts of the ceiling are covered with short soda-straw stalactites, the walls are coated with creamy white flowstone, and there are even nests of cave pearls. The overlying concrete pavements are the likely source of the calcium minerals in the cave formations. The acoustics of the tunnel are such that the dripping water mimics the sound of muffled speech, a creepy circumstance for Greg's friend, who refused to explore alone!

While it was fascinating to walk under the downtown area, Greg did not find any stream of water that was unambiguously the actual Jordan Creek. The tunnels are not open to the general public.

Directions:	The Jordan Creek historical marker is bolted to the sidewall of the building at 425 West Third Street in downtown Red Wing. You can hear rushing water in nearby manholes.
Seasons/Hours:	Year-round.
Length:	N/A
Precautions:	Visiting the storm drains is not recommended owing to the dangers.
Amenities:	Check out nearby Marie's Underground Grill & Tap House, 217 Plum St., in the basement of the former Red Wing Armory building.
Information:	Goodhue County Historical Society, 1166 Oak St., Red Wing, MN 55066. Phone: 651-388-6024. Website: http://goodhuecountyhistory.org.

LE DUC HOUSE
DAKOTA COUNTY

The river town of Hastings is underlain by a vast natural labyrinth, according to local legend. Some residents believe in a smaller version of this labyrinth, one that connects Miles Cave, a natural cave in the bluffs of the Vermillion River, with the basement of the Le Duc House, a nearby mansion. But the cave is much shorter than most people realize, about 500 feet of dusty crawling passages, none of which reach even so far as the mansion. The caver and fossil collector Tim Stenerson of Red Wing, who contributed so much to documenting and mapping the smaller, seldom visited caves of Minnesota before his untimely demise, spent many happy hours digging out the crawlways of Miles Cave, always hoping for that elusive connection to the Le Duc House.

The historic Le Duc House, now operated by the Dakota County Historical Society, was named after William Gates Le Duc (1823–1917), the U.S. commissioner of agriculture under President Rutherford B. Hayes. During the candlelight Halloween tours of this mansion, a guide will often point to a patch on the basement wall as marking the location where the supposed connection to Miles Cave used to be.

Miles Cave is rumored to honeycomb the ground beneath the town of Hastings. (Photo: Tim Stenerson)

Miles Cave, on the other hand, was reportedly named after General Nelson Appleton Miles (1839–1925) of Civil War fame. According to one witness, "I was told by early settlers that the Indians often used this cave. Also, that it was named after General Miles who used the cave in his dealing with the Indians, and his successful ventures in Indian outbreaks. This is how it finally received its name." It often seems that General Miles' association with the Battle of Wounded Knee provides an apt metaphor for anyone trying to crawl through this rocky floored cave!

Although heavily visited by local residents, Miles Cave is actually on private land owned by the CONAGRA flour mills, so the following narrative is for descriptive purposes only.

Situated in the Oneota dolomite layer downstream from the falls of the Vermillion River, the largest room in Miles Cave is the so called Party Room, which with its natural rock benches served as a cave classroom for Greg's Speleology 101 course offered through the University of Minnesota years ago. The room is adorned with flowstone and cave pearls, which form where sand or gravel gets coated with calcite and becomes rounded because of the impact of falling water droplets. Centrally located, crawlways radiate out from this room like the spokes of a wheel, one of them leading to an exit in the open fields above. During especially cold winters, the cave grows a forest of club-shaped ice stalagmites.

So much for the eastern half of Miles Cave. The western half is much tougher to get to, either by a dangerous cliff climb or through the Carcass Crawl, where an unconfirmed date of 1848, discovered by Stenerson, has been carved in the ceiling. The chilly pools of water, raccoon carcasses and scats that carpet this lengthy claustrophobic crawlway are powerful dissuasions to those passing over to the other side. Not to mention the prospect of encountering a disgruntled raccoon somewhere in between! In addition to raccoons, Greg has observed house cats prowling the forlorn passages of Miles Cave. Perhaps they have found the secret connection to the mansion?

Directions:	Le Duc House, 1629 Vermillion St., Hastings, MN 55033
Seasons/Hours:	Open for tours May through October. Wednesday, Friday – Sunday 10 a.m. to 4 p.m.; Thursday 10 a.m. to 8 p.m. Candlelight tours of the cellars are held at Halloween. Fee. Available year-round by reservation for group tours and private events.
Length:	N/A
Precautions:	The cave is on private land and getting to it involves scaling cliffs; not recommended.
Amenities:	Gift shop.
Information:	Phone: 651-437-7055. Website: www.dakotahistory.org

THE CRUSHED MAN OF LEE MILL CAVE

Lee Mill Cave, with its magnificent view of Spring Lake, formed in the Oneota dolomite of Schaar's Bluff, near Hastings, Minnesota. Excavated by the Science Museum of Minnesota in the early 1950s, archeologists found human bones crushed beneath a boulder, along with pottery, and large numbers of fish bones. The cave today is littered with raccoon bones and scat. It features a natural chimney that conducts smoke upward through a crevice in the rock, which perhaps was another reason why early peoples favored the cave. Past a constriction in the cave passage, there's another, smaller room, a sort of inner sanctum, warm in winter and cool in summer. Any visitor with a tinge of arachnophobia will find this small back room, crawling with large spiders, a bit disturbing. The gristmill that gave the cave its name was owned by the Lee family, and the land remains private property.

Greg Brick in Lee Mill Cave. (Photo: Tony Andrea)

CRYSTAL LAKE CAVE
DUBUQUE COUNTY, IOWA

Miners hoping to find a vein of lead ore found instead Crystal Lake Cave in 1868. James Rice and his crew sank a shaft 40 feet through the limestone, and Rice, coming on a crevice, explored 700 feet of a natural passage, finally reaching a large cave system. The first adventurous visitors to the then-called Rice's Cave entered via a bucket lowered by rope into a shaft.

What they saw was part of the same mineral wonderland you can see today, albeit with more layers of mud and rock fragments on the floor. The cave system, which extends for several miles, lies 75 to 100 feet underneath the rolling hills; some passages remain inaccessible to most visitors. But what is accessible is breathtaking.

Developed and officially opened to the public in 1932, Crystal Lake Cave has preserved many beautiful formations, including stalactites, stalagmites, helictites, and several rare anthodites, or cave flowers. These white clusters somewhat resemble a dahlia, with narrow petals radiating from the center. They actually are crystals of aragonite, a mineral made up of calcium carbonate and in some ways similar to calcite.

Today you enter nature's showroom by descending 28 concrete steps. (Though steep, this stairway represents a considerable improvement over a rope and bucket!) The walkway at the base of the stairway is also paved, leading you through the passages as surely as Dorothy's Yellow Brick Road.

Your Emerald City, however, is Crystal Lake. Lying at the end of a long passage, the lake extends 28 feet into the main passage ahead of you. It is three to four feet wide and two-and-a-half feet deep. But its real attraction hangs overhead and on the glassy surface where the reflections of a multitude of stalactites shimmer in the lamplight.

St. Peter's Dome at Crystal Lake Cave. (Photo: Crystal Lake Cave)

All the rooms that break up the narrow passageways seem relatively small; however, couples have held marriage ceremonies in the Chapel Room. And the undeveloped portion of the cave includes two large rooms, including the Flat Room, which extends 90 feet through the limestone. Many of the rooms and shoulder-brushing passages are festooned with soda straws and other stalactites and furnished with stalagmites.

Named years ago, the wonderful formations really look like their names. You can readily see the Tyrannosaurus Rex, for example, not to mention all the animals of Noah's Ark, in Imagination Alley. And, the Swiss cottage roof with its overhanging straw or icicles belongs in a stage version of Heidi. The cave offers a kind of field of dreams, where every room stirs a memory—here a petrified forest, there a Disney cartoon character, and over there, perhaps, an evocative movie hero from childhood.

Breaking the dream, an earthenware jug rests in a small niche in the wall, a leftover from the 19th century. And a sign proclaims, "$50 fine for breaking or destroying crystals." You're glad to see this effort at cave conservation but wish the intrusive reminder was unnecessary.

Too soon, you reach Tall Man's Misery, the five-foot-three-inch-high final passage. The 45-minute tour seemed like a quarter of an hour. You exit by climbing another set of steps and walk outside perhaps 200 feet south of the entrance.

After a picnic on the Crystal Lake Cave grounds, you may want to check out the nearby **Mines of Spain State Recreation Area** (http://www.minesofspain.org). Though not offering any opportunities to venture underground, it honors the first European to mine the area. Julien Dubuque received a land grant from the governor of Spain in New Orleans in 1796. The grant gave Dubuque the right to work the area, to be called "Mines of Spain." The recreation area features the 1897 Julien Dubuque Monument (an interpretive center), picnicking, and 21 miles of hiking trails. To reach the Mines of Spain, drive north toward Dubuque on U.S. 52 from Crystal Lake Cave and watch for signs.

Directions:	Located five miles south of Dubuque, left off of U.S. 52. Watch for signs.
Seasons/Hours:	Open daily from 9 a.m. to 6 p.m. from Memorial Day to mid-August. See website or call for reduced spring and fall hours, from early May through October. Fee. Last tour leaves one hour before closing. Phone to learn more about the one-and-a-half hour wild cave tour.
Length:	Approximately three quarters of a mile; tour lasts 45 minutes.
Precautions:	Light jacket recommended; temperature is 50 degrees. Lowest passage measures five feet, three inches in height.
Amenities:	Picnicking, shelters, restrooms, gem-mining sluice for kids.
Information:	Crystal Lake Cave, 6684 Crystal Lake Cave Dr., Dubuque, IA 52003. Phone: 319-556-6451. Website: www.crystallakecave.com.

SPOOK CAVE
CLAYTON COUNTY, IOWA

The story of Spook Cave began with a mysterious spring at the base of a 90-foot bluff along Bloody Run Creek, near McGregor, Iowa. Strange sounds emanating from a small opening at the spring resulted in the name Spook Hole. Gerald Mielke, a local resident who had experience developing show caves in the Upper Midwest, saw commercial possibilities. In 1953, he blasted into the bluff and discovered a sizable cave—along with flowing water inside the cave that was causing the spooky sounds. He blasted open a separate work entrance in a side ravine and built a ramp down into the cave. He spent the next two years removing tons of rock and mud. The cave opened for business in 1955 and remains the only underground boat tour in Iowa.

Spooky reflections at the boat entrance to Spook Cave. (Photo: Greg Brick)

The ticket stand and gift shop for the cave tour is Spook Cave Lodge, created by grafting two local schoolhouses together. It contains a wall of historic photos documenting the commercial development of Spook Cave. The tour departs from the dock in one of Mielke's original seven custom-built aluminum boats, with a quarter-horsepower electric trolling motor at the bow. The boat floats across the millpond, under the footbridge, and into the cave entrance. The milldam maintains a sufficient depth of water to float boats in the cave stream, known as the Spook River. In winter, the millpond is drained away and Spook River falls to its natural, shin depth. Mielke built the nearby Old Mill for the purpose of generating electricity for the cave lights, but the waterwheel didn't produce much power, so the mill became merely ornamental. The millpond flows into Bloody Run Creek, which empties into the Mississippi River at Marquette. The creek got its name because this area was a favorite hunting ground for soldiers from Fort Crawford, across the Mississippi in Prairie du Chien, Wisconsin. They washed animal hides in the creek.

Spook Cave developed in the Galena limestone and the guides will tell you that it's 750,000 years old. The Spook River flows on top of the underlying, impervious, Decorah

Shale. The cave is an elliptical rock tube with domes at intervals. The cave drains 13 square miles of land surface, discharging 1.75 million gallons of water per day. The water becomes turbid after a heavy rainfall or during spring snowmelt, carrying agricultural sediments into the cave and depositing them in the boat channel. As a result, the cave needs to be dredged each winter, in the off-season.

The roof of the cave is very low in places, requiring you to lean forward, almost to the point of placing your head in your lap. The first such place, just inside the entrance, is called Lover's Lane for this very reason—you may need to get cozy with the fellow occupants of your boat! The handles affixed to the cave walls aid navigation in the tight spots but they are for the use of the guide only. Hands should be kept inside the boats because they could easily get crushed when the boats bang against the walls. Deep grooves are worn in the rock from the constant rubbing of the boats.

As your boat passes from the artificial, blasted section into the natural cave proper, a side passage, called Mielke's Crawl, comes into view on the right side. The roof suddenly lifts and you float into the Big Room, the largest room in the cave, 30 feet high and 110 feet long. Look for the trapdoor in the ceiling, the location of the former upper entrance to the cave. Although it was originally thought that tours could be brought in this way, the opening created a strong air current that began drying out the cave formations, so it was sealed. The left wall of the Big Room, composed of jumbled rock and mud, is called The Landslide.

Upstream from the Big Room, on the right side, is the zebra-striped Flowstone Wall. The white stripes of pure calcite alternate with black stripes containing manganese compounds. On the ceiling here you'll see the bottom of a sinkhole. While it's dry most of the time, this hole showers water after a heavy rainfall. On the left, you pass the illuminated Wishing Well, which is ornamented with formations that broke off during commercial development of the cave.

After rounding the S-bend of Spook Cave, the tour guide is sure to tell the story of Old Joe for comic relief. In one version of this yarn, Old Joe is a solo cave explorer and, in another, a light-bulb changer (whether there were all that many bulbs to change is beside the point). Old Joe didn't return one day, so they say, and his boat was found capsized in Spook River. Scratch marks were found on the chocolate-colored mud banks where he tried to save himself—in a foot or two of water, at best—a feat that's roughly comparable to drowning in a bathtub. His body was never found, but the guide will point out his surprisingly new hat resting on the mud bank. You may purchase a similar hat in the gift shop!

You now enter a low-ceilinged, elliptical rock tube. About midway through, as you approach the Dog Tooth Formations, you again have to hunker down in the boat. This cluster of tooth-like stalactites, which includes a "hairy" stalactite, is protected by a screen on the ceiling. Just after that, the Frozen Waterfall will give you a chance to tip your head back in the other direction as you peer up into the 35-foot-high dome. The Frozen Waterfall, an undulating, flowstone-covered wall, was the high point of the original cave tours in the 1950s and the place where boats turned around to head back for the entrance. It wasn't until years later that the next segment, Formation Alley, was opened for business. Formation Alley has a well-developed "life-line" (as guides are fond of calling joints in the roof of the cave) crammed with stalactites and ribbed with bacon strips—a kind of cave formation that resembles thin strips of bacon. During a winter trip into Spook Cave, kindly arranged by the owner, this was seen to be a life-line in a different sense—it harbored numerous moisture-loving tricolored bats.

Near the end of the tour you enter the Dome Room, 40 feet high and 120 feet below the surface. Illuminated by your guide's flashlight, you'll see the Rock of Ages, a waterproof enamel painting on glass, balanced on a high rock ledge in the cave, depicting a woman clasping a stone cross in a surging tide. The passage is wide, leaving ample room for the boats to turn around. Just beyond is Half Mile Bar, spanning the passage like a chin -up bar. Despite its name, it's only about a quarter of a mile from the entrance. Beyond the bar there's a dam across the stream, designed to stop agricultural sediments from filling in the boat channels, which they seem to do anyway.

Beyond the dam, which marks the end of the commercial tour, the passage is filled with huge rock slabs and ends at the Sump Room, which contains the deepest water in the cave. Cave divers have explored a thousand feet beyond this sump through water-filled passages.

On the grounds outside Spook Cave, just downstream from the boat entrance, you'll see a spring called Beulah Falls pouring from the cliff. This little spring cave will give you some idea of what the original Spook Hole may have looked like. It was named after Beulah, a small railroad town located where the campground is today. The town was swept way in a flood many years ago.

For more Iowa sites, see Greg Brick's 2004 book, *Iowa Underground*.

Directions:	Located west of McGregor, IA. From U.S. 18, turn north on Spook Cave Road, follow the signs.
Seasons/Hours:	Open daily 9 a.m. to 5:30 p.m., May through October. Fee.
Length:	The boat tour, through 1,000 feet of passage, lasts 40 minutes.
Precautions:	"Always 47 degrees" the brochure says, so wear a sweater or jacket.
Amenities:	Advertises "93 Acres of Fun," including gift shop, game room, cabin rental, campground, playground, picnicking, hiking, swimming, and trout stream.
Information:	Spook Cave and Campground, 13299 Spook Cave Rd., McGregor, IA 52157. Phone: 563-873-2144. Website: www.spookcave.com.

SOUTHWESTERN MINNESOTA

Drive from east to west in southern Minnesota and the landscape changes dramatically—from the steep stream valleys, rugged bluffs, and sinkholes to drier, more level terrain. The last glaciers of the most recent Ice Age advanced across the southwestern part of the state while leaving the Southeast untouched. More than 10,000 years later, this demarcation still shows above and beneath the surface of southern Minnesota.

When the last glacier retreated from southwest Minnesota, it left a thick layer of soil, stones, rock and other debris, called till. Eventually the glacial till supported the vast prairies that European explorers found here and that 19th-century pioneer farmers cultivated.

Depending on your route, you may encounter a highland in Minnesota's southwest corner, edged with a cliff—called an escarpment or scarp—visible in sections of the east-facing side. This knife-edged upland separates the James River and Minnesota River basins and is the Mississippi River-Missouri River watershed.

Southwestern Minnesota is also a drier part of the state without the soluble rocks found in the more humid southeast, so its underground attractions have a different aspect. The caves tend to be artificial, or, if natural, created by earth movements.

JORDAN BREWERY CAVE
SCOTT COUNTY

In 2019, the historic Jordan brewery complex, complete with a 625-foot cooling cave, was on the market for $1.35 million. Located southwest of the Twin Cities off U.S. 169, the complex, which now includes apartments and two historic homes, began as a brewery in 1861. Operated by Sebastian Gehring, the brewery must have been successful, because in 1879 Fred Heiland built another brewery adjacent to the first.

Various owners operated a brewery at the eight-acre site for many decades, except during Prohibition, when it became a fish hatchery, with eggs kept in the manmade cave. Although the complex was again used as a brewery after the repeal of Prohibition, business waned for various reasons and in 1954 a fire did serious damage.

Thanks largely to the efforts of Gail Andersen—who first bought the property in 1972, sold it, and then repurchased and renovated it—and her granddaughter Barb

Kochlin, who bought the property in 2011, the brewery complex exists today. It's on the National Register of Historic Places.

The brewery cave was carved into the Jordan Sandstone bluff behind the building. A survey completed by the Minnesota Speleological Survey in 1981 documented an entry passage, partially water-filled storage areas, plank bridge, stone wall, and areas of collapsed debris. The unique feature of the cave from a geological standpoint is that some of the passages are at the contact of the sandstone with the glacial drift, such that the walls and floor are sandstone and the ceiling is drift.

Directions:	415 Broadway Street South, Jordan, MN.
Seasons/Hours:	Year-round.
Length:	The cave (private) is 625 feet long.
Precautions:	N/A
Amenities:	N/A
Information:	Jordan Area Chamber of Commerce, www.jordanchamber.org

JESSE JAMES CAVE
NICOLLET COUNTY

While Jesse James and his gang robbed banks, others made a living off their exploits, taking the money right back to the bank. Along U.S. 169 south of St. Peter, Minnesota, several holes in the sandstone outcrops alongside the highway are visible. Although the caves are artificial, according to some sources, they were merely enlarged from natural caves formed by groundwater flowing through the Jordan Sandstone. In any case, they are the only remnants of a much larger cave system developed by Charlie Meyer, who bought the property back in 1930.

Working in the Kasota stone quarries across the Minnesota River, Meyer was an expert sculptor. He lovingly took the time to carve elaborate historical figures and painted them, creating a nice show cave that he called Jesse James Cave or sometimes Seven Caves. The signature theme was that the cave was used as a hideout by Jesse James on his way to robbing the Northfield Bank in 1876. Although these stories had circulated for years, a surviving member of the gang later denied the use of caves for this purpose, pointing out that caves just as easily became traps for those pursued.

As usual for sandstone caves, with their lack of natural ornamentation such as stalactites and stalagmites found in limestone caves, something had to be done in the way of artificial embellishment to compensate. Whether the gang had used the cave or not, Meyer commemorated each member with a carved head in the wall. But Meyer's life

Entrance to one of the remaining Jesses James caves along U.S. Highway 169. (Photo: Greg Brick)

Looking out from one of the Jesse James caves. (Photo: Greg Brick)

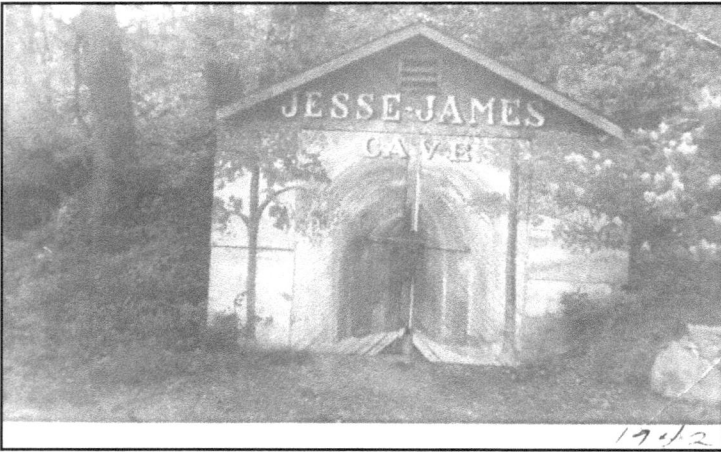

The former entrance building of Jesse James Cave. (Photo: Nicollet County Historical Society)

work was wiped out when U.S. 169 came through in 1954, destroying four of the original seven caves.

Whether or not Seven Caves afforded refuge to the Jesse James gang, they certainly afforded refuge to hibernating bats. During the great Armistice Day blizzard of 1941, many bats could not relocate the cave entrances, which had drifted shut with snow, and perished miserably outside. But it was a good day for the local cats, which carried away the frozen bat carcasses.

What's left for public viewing from the highway right of way is a sorry remnant of the much larger cave system. The entrances may be seen near the mouth of a ravine. When cavers visited years later, they found long stooping passages flooded with cold water. There may be some traces of the stony face on the walls, but it's easy to fool yourself in the dark!

Directions:	Entrances to two of the caves are still visible along the right-of-way of U.S. 169 at mile 63.7, south of St. Peter, MN. It's safest to park on Freeman Drive and walk south along the highway for one quarter mile, until coming to the deep ravine where two remaining caves are found. The caves are located at UTMs 421090 E, 4905403 N.
Seasons/Hours:	Year-round.
Length:	60 and 150 feet.
Precautions:	The caves are along the public right-of-way of a busy highway. Park safely and walk as far off the road as possible. Please respect private property, as caves more remote from the highway are not publicly accessible.
Amenities:	None.
Information:	Nicollet County Historical Society, 1851 Minnesota Ave. N, St. Peter, MN 56082. Phone: 507-934-2160. Website: www.nchsmn.org

JESSE JAMES CAVE, BLUE MOUNDS STATE PARK
ROCK COUNTY

A curving cliff of pink and purple Sioux quartzite forms the mile-and-a-half backbone of Blue Mounds State Park and contains talus, fissures, and crevices that, according to legend, once harbored the James gang and were depicted as such on old postcards. You can explore these deep fissures along the east side of the 90-foot cliff on the Upper and Lower Cliffline Trails.

The quartzite was laid down more than one and a half billion years ago on the bottom of an ancient sea. First, sand was deposited on the sea floor and with the weight of accumulation and water turned into sandstone. Then, time, heat, and chemical reactions converted the sandstone into quartzite. Much later, glaciers gouged into the quartzite when loose rocks were dragged across the bedrock. The pink to purple colors in the quartzite are due to the iron oxide minerals. You can see evidence of this geological history, as well as remnants of quarrying, along the park cliff.

At the cliff's southern end a 1,250-foot line of rocks runs in an east-west direction. At the spring and fall equinoxes the sun at sunrise and sunset lines up with this stone alignment. Visitors sometimes hike to view the line of stones and to ponder its origin.

Most of the park's 1,500 prairie and grassland acres lie on top of the quartzite outcrop. This is where the buffalo roam. Visitors are most likely to spot the bison in their

Jesse James Cave at Blue Mounds State Park, another reported Jesse James Gang hideout.
(Postcard from the Gordon Smith Collection)

pasture early in the morning or on a bison tour. Coyotes and deer can also be seen in the park; prickly pear cacti bloom on the outcrops in summer.

Directions:	From Luverne, go north 4 miles on U.S. 75. Turn east on County Highway 20 and go 1 mile to the park entrance, which leads to the park office, where you can pick up a park map. To reach the talus caves, return to U.S. 75 and drive south. Turn left (east) on County Highway 8 and drive to a parking area on the left.
Seasons/Hours:	Open year-round; however, some facilities are closed in winter. Minnesota State Parks permit required.
Length:	The walk from the parking area to the cliffs and back is less than 2 miles.
Precautions:	The hike to the cliffs is moderately challenging, with some uphill sections.
Amenities:	Camping, rock climbing, hiking, biking, snowmobiling, lake fishing, canoeing, bison herd, gift shop, historic Fredrick Manfred Home.
Information:	1410 161st St., Luverne, MN 56156. Phone: 507- 283-1307. Website: www.dnr.state.mn.us/state_parks/index.html and use the Park Finder.

HINKLY HOUSE, BLUE MOUNDS STATE PARK
ROCK COUNTY

The tunnel at the Hinkly House served as a storage cellar for vegetables, grape juice—and dynamite. A Luverne founding father, R. B. Hinkly, built his Victorian home out of Sioux

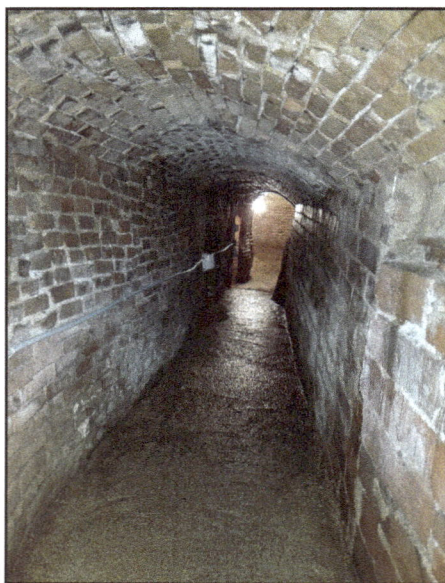

quartzite mined from his quarry where dynamite was frequently used to break down the rock. His quarry is now part of Blue Mounds State Park.

During the summer the Rock County Historical Society opens Hinkly's three-story home to visitors three days a week. Approximately five and a half feet in height, the tunnel runs about 35 feet to its

Hinkly House Tunnel led to a dynamite storage room at some distance from the home. (Photo: Luverne Area Chamber of Commerce)

The Hinkly House dynamite storeroom today holds old bottles of grape juice. (Photo: Luverne Area Chamber of Commerce)

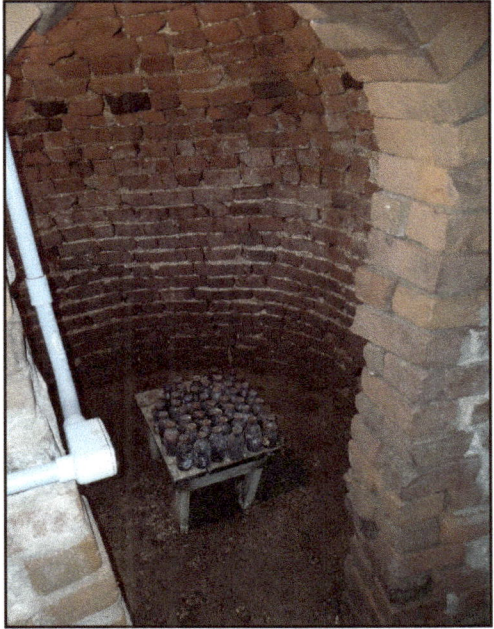

The Hinkly House dynamite storeroom today holds old bottles of grape juice. (Photo: Luverne Area Chamber of Commerce)

end beneath the backyard, with the dynamite room at the terminal point. The tunnel turns before reaching the dynamite room, so that if there were an explosion it wouldn't go around the bend and into the house. As an added safety measure, Hinkly stored the caps in the library.

Alcoves in the tunnel's first section held jars and vegetables, as well as little bottles of Hinkly's homemade grape juice, apparently a favorite treat. Today a number of these bottles are on display in the dynamite room.

Directions:	From U.S. 75 in Luverne, drive two blocks north from intersection with Main Street. Turn east (right) and continue two blocks to 217 Freeman Avenue.
Seasons/Hours:	Hinkly House is open June through August, Tuesday, Thursday, and Saturday, 2 to 4 p.m. Free-will donations.
Length:	N/A
Precautions:	Hinkly House is accessible.
Amenities:	Luverne offers restaurants, two museums, and the Brandenburg Gallery where you can view and purchase the work of National Geographic photographer Jim Brandenburg.
Information:	217 Freeman Ave, Luverne, MN 56156. Phone: 507-449-2115. Website: www.exploreminnesota.com/things-to-do/2728/hinkly-house.

PIPESTONE NATIONAL MONUMENT
PIPESTONE COUNTY

Long ago, the Great Spirit took the form of a large bird and called together all the Native American tribes at a wall of rock. He broke off a piece of the red rock, formed it into a pipe, and smoked it. The Great Spirit told the tribes that this stone was their flesh and that they must use it only for sacred objects, according to an account recorded by artist George Catlin in 1836.

Even today, the 301-acre Pipestone National Monument offers a spiritual experience as much as a cultural or geologic point of interest. While anyone can walk through the historic quarry, only Native Americans can obtain a permit to quarry and sell the soft, red pipestone (also called catlinite after the artist who brought the stone to the attention of 19th-century scientists). Walking the Circle Trail through the quarry, you will see ancient red walls, evidence of recent digging, and occasionally small packets of bright offerings tied to overhanging tree limbs.

Native American tribes from as far away as the American Southwest and Canada come to excavate the pipestone in the summer and fall, when water does not collect in the ditches they dig by hand. The quarriers shovel away any soil and then break up the 10-to-15-foot layer of hard quartzite that lies over the softer pipestone. Both the quartzite and the pipestone formed about 1.4 billion years ago; however, quartzite is the result of great pressure and heat applied to sand, while pipestone formed from ancient clay. Its red color results from oxidation of trace amounts of iron.

The pipestone layer is typically only 14 to 18 inches thick, with only a few inches suitable for carving pipes or other artwork. The quarriers break the pipestone along natural fissures in the rock and take it out in sheets from one to three inches thick. They then cut the sheets into blocks appropriate for carving. Pipestone is about as hard as your fingernail and can be easily carved using a steel knife blade. Only the pipe bowl is crafted from pipestone; the pipe stem is made from ash or other hardwoods.

Walking along the Circle Trail, you pass several active and inactive pits, a high wall of roughly eroded red quartzite, Pipestone Creek, and Winnewissa Falls, which rushes over the red wall. You also travel briefly "underground" beneath a red rock archway. Markers note natural and cultural phenomena:

- Wind and water erosion has created the Old Stone Face formation in the wall near the falls.

- Leaping Rock was the site where young warriors would prove their valor by leaping into a chasm and placing an arrow in a crevice.

- One rock face bears the carved initials of members of the 1838 expedition to the Upper Mississippi region led by Joseph N. Nicollet.

- Old tribal shamans believed they heard voices issuing from the stone lips of the Oracle formation.

- A remnant of a tallgrass prairie changes color with the seasons as different grass and flower species bloom.

Years ago, tribes came together to mine the sacred pipestone in peaceful accord. They still come in an annual pilgrimage to follow an ancient custom of careful craft and art. Today, that craft and art is open for all to witness and appreciate. Displays in the visitor center showcase selected pipe bowls, stems, petroglyphs, and other artwork.

Directions:	Follow I-90 west to U.S. 75 north through the town of Pipestone. Signs will lead you to the monument.
Seasons/Hours:	Open daily, 8:30 a.m. to 5 p.m., except Thanksgiving, Christmas, and New Year's Day.
Length:	The paved Circle Trail through the quarry is about 0.75 mile in length. Allow 45 minutes to take in the quarry and the marked points of interest.
Precautions:	Although the Monument is open year-round, parts of the Circle Trail may be closed in winter, due to snow and ice accumulation. Taking rock samples is prohibited. Motorized vehicles and bicycles are not allowed on the trail. While the trail is negotiable by wheelchair and wheelchairs are available at the visitor center, wheelchair users will require assistance to navigate some sections of the trail. While motorized scooters are allowed on the trail, users need to beware of sharp turns, steep slopes, and drop-offs. Not all scooters will fit on the trail's cliff-line section.
Amenities:	Visitor Center with gift shop, restrooms, and paved trail to the quarry. The city of Pipestone offers an historic district to explore, restaurants, shops, and lodging.
Information:	Pipestone National Monument, 36 Reservation Ave., Pipestone, MN 56164-1269. Phone: 507-825-5464 ext. 214. Website: www.nps.gov/pipe/index.htm.

JEFFERS PETROGLYPHS

About 5,000 images are carved into an outcrop of Sioux quartzite about 65 miles east of Pipestone National Monument. The 7,000-year-old carvings depict humans, animals, and weapons, as well as undecipherable shapes.

Maintained by the Minnesota Historical Society, this site is open from approximately Memorial Day to Labor Day, plus for group tours with reservations at other times. Fee.

To reach the Jeffers Petroglyphs, follow U.S. 71 to County Highway 10. Follow 10 east for 3.0 miles to County Highway 2. Turn right on 2 and drive 1.0 mile to the visitor center. Website: http://www.mnhs.org/jefferspetroglyphs.

Greg taking notes in an urban cave. (Photo from the Greg Brick Collection)

Coldwater Spring with its stone-walled reservoir was the entrance to the Dakota underworld. (Photo: Greg Brick)

MINNEAPOLIS-ST. PAUL
METROPOLITAN AREA

T he chief population center of Minnesota, embracing the capital city of St. Paul, with its larger twin, Minneapolis, is situated in the second-most cave-rich region of the state, after the southeastern karst counties, especially Fillmore County. However, the most abundant sort of cave here is the artificial sandstone cave, and even the natural caves, excavated by flowing water, differ markedly from the limestone caves of the Southeast, which formed by dissolution from slightly acidic groundwater. Yet early cave history in Minnesota is most closely bound up with the Twin Cities. And in terms of tunnels, no other place in the state can remotely compare to its hundreds of miles of utility tunnels. See Greg Brick's 2009 book *Subterranean Twin Cities* for more in-depth coverage of this aspect.

COLDWATER SPRING
MISSISSIPPI NATIONAL RIVER & RECREATION AREA
HENNEPIN COUNTY

"Camp Coldwater," according to historians Bruce White and Dean Lindberg, "was the first settlement of European-Americans in Minnesota that was not primarily a fur trading post, fort, or mission....The site...was the location of many 'firsts' in Minnesota history, a good reason to call it the birthplace of Minnesota." Indeed, Camp Coldwater, in Minneapolis, has been called Minnesota's Plymouth Rock. The spring that gave its name to the encampment gushes from the Platteville Limestone at 60 gallons per minute. As early as 1819, the spring supplied early Fort Snelling with cold drinking water—a welcome alternative to the warm, turbid waters of the nearby Mississippi River. A stone-walled reservoir was built in 1880, impounding the waters. In later years, this reservoir served as a trout pond on what became the U.S. Bureau of Mines' property, now part of the Coldwater Spring Unit managed by the National Park Service. The trout devoured the amphipods that swarmed in the pool, one of their favorite foods. But the spring is also sacred to Native Americans, for a special reason.

Unktahe (many variant spellings of the word exist) was the Dakota god of water and the underworld. Historically, the Camp Coldwater spring was associated with Unktahe, who was often visualized as a fish or serpent. Mary Henderson Eastman wrote in 1849, "Unktehi, the god of the waters, is much reverenced by the Dahcotahs. Morgan's Bluff, near Fort Snelling, is called 'God's house' by the Dahcotahs; they say it's the residence of Unktehi, and under the hill is a subterranean passage, through which they say the water-god passes when he enters the St. Peter's [Minnesota River]. He is said to be as large as a white man's house."

James Owen Dorsey, in his "Study of Siouan Cults" for the U.S. Bureau of Ethnology wrote about "The Unktehi, or Subaquatic and Subterranean Powers," in 1894:

> *The gods of this name, for there are many, are the most powerful of all. In their external form they are said to resemble the ox, only they are of immense proportions. They can extend their horns and tails so as to reach the skies. These are the organs of their power. According to one account the Unktehi inhabit all deep waters, and especially all great waterfalls. Two hundred and eleven years ago, when Hennepin and Du Luth saw the Falls of St. Anthony together, there were some buffalo robes hanging there as sacrifices to the Unktehi of the place....It is believed that one of these gods dwells under the Falls of St. Anthony, in a den of great dimensions, which is constructed of iron.*

More than a century later, Gary Cavender, a local Native American spiritual leader, filed an affidavit in the Highway 55 case, in which the rerouting of the nearby highway threatened Coldwater Spring. In 1998, he stated:

> *The Camp Coldwater spring is a sacred spring. Its flow should not be stopped or disturbed. If the flow is disturbed, it cannot be restored. Also, if its source is disturbed, that disturbs the whole cycle or the flow. The spring is the dwelling place of the undergods and is near the center of the Earth. The spring is part of the cycle of life. The underground stream from the spring to the Mississippi River must remain open to allow the gods to enter the river through the passageway. The spring is the site of our creation myth (or Garden of Eden) and the beginning of Indian existence on Earth. Our underwater god (Unektehs) lives in the spring. The sacredness of the spring is evident by the fact that it never freezes over, and it is always possible to see activity under the surface of the water.*

Coldwater Spring gushes from the stone spring house at a corner of the reservoir. Upon closer inspection of its conduits, geologists only recently discovered that it was a triple spring whose annual chemical cycles of salt and temperature reflected the city's metabolism. But the only suggestion of Unktahe's Cave is when the stone-walled pool becomes heavily overgrown with aquatic vegetation. Staring into the murky depths, it almost appears as if some creature is stirring down there, among the waving fronds.

Directions:	From Hiawatha Avenue (State Hwy 55), turn east onto 54th St. Take an immediate right and drive to the end of the street, where you'll find a gravel parking lot. Hike the south-going trail one quarter mile until arriving at the walled spring pool.
Seasons/Hours:	Year-round, 6 a.m. to 10 p.m.
Length:	Quarter-mile gravel hiking trail to the spring.
Precautions:	The restored prairie can be blazing hot in summer; carry water.
Amenities:	Hiking trails lead down to the Mississippi River and connect with those from Fort Snelling.
Information:	National Park Service, 111 Kellogg Blvd. E, Suite 105, St. Paul, MN 55101. Phone: 651-293-0200. Website: www.nps.gov/miss/planyourvisit/coldwater.htm.

FORT SNELLING NATIONAL HISTORIC LANDMARK
HENNEPIN COUNTY

"History Under the Floor Boards" provides an archaeologist's view of the excavation in the "basement" of the historic Fort Snelling Officers' Quarters. The multimedia exhibit pinpoints the bases of two limestone fireplaces, walls, and other features revealing how the building evolved since its original construction in the 1820s. An 11-minute video explains how archaeologists dated the addition of the second fireplace to the 1840s and analyzed more than 28,000 fragments of dishes, utensils, marbles, and other artifacts to

Rumor has it that this abandoned utility tunnel was dug by enemy forces trying to capture Fort Snelling!
(Photo: Greg Brick)

77

piece together a picture of life in the Fort. Young would-be archaeologists are usually intrigued by several vignettes that depict real-life exchanges among family members living in these quarters in the 1820s. (Oops! There goes another marble, fork, or hairpin—lost through a crack in the floorboards.)

Children appreciate the many interpretive programs at the Fort, especially the soldiers firing muskets and cannons, blacksmith shop, and fur trade encampment. A small regiment of interpretative guides in period costumes takes visitors back to life on the Minnesota frontier. Guided tours have sometimes included Quarry Island (also known as Wakan or High Rock Island), in the adjoining Fort Snelling State Park, to see clues left by a 19th-century quarrying operation. The "island" is connected to the shore by a long causeway.

Adults are often more impressed by the scope of the Fort's reconstruction, which began in the 1950s. At the time, only four buildings remained standing (officers' quarters, round tower, hexagonal tower, and the commandant's house). The restoration took 15 years and $4 million.

Minor caving opportunities once existed outside the walls of Fort Snelling. Around 1900 there were newspaper reports of Boy Scout trips to the Fort's root cellars, but diligent search in recent times has revealed only collapsed structures in the woods. And heading north along the hiking trail that runs alongside the base of the Fort you'll find the entrance to a 60-foot-long stooping height tunnel in the sandstone. Rumor has it that it was dug by enemy forces trying to capture the Fort but it's really just an abandoned utility vault.

The longtime survival of the Fort depended considerably on the reliable water source of the Coldwater Spring at Camp Coldwater, located upstream on the west bank of the Mississippi River. (See COLDWATER SPRING)

Directions:	Located near the Minneapolis-St. Paul International Airport near the intersection of State Highway 5 and State Highway 55, above the confluence of the Minnesota and Mississippi rivers.
Seasons/Hours:	Memorial Day through Labor Day, Tuesday through Friday, 10 a.m. to 4 p.m. and Saturday and Sunday, 10 a.m. to 5 p.m. Closed: November through April. Fee.
Length:	Allow several hours to explore the historic fort.
Precautions:	The journey from the parking area to the fort is an easy 10-minute walk.
Amenities:	The Fort comprises more than a dozen restored buildings, including a hospital, officers' quarters, barracks, schoolhouse, blacksmith and harness shops. The History Center features exhibits and films, and hiking trails run along the Mississippi River below the Fort.
Information:	200 Tower Ave., St. Paul, MN 55111. Phone: 612-726-1171. Website: www.mnhs.org/fortsnelling.

SUGAR CAVE AND THE SAND PAINTER

Nestled on the banks of Minnehaha Creek, near Godfrey's Point, where it joins the Mississippi River, in Minneapolis, is Sugar Cave, partly natural and partly artificial, which has a colorful history disproportionate to its very modest dimensions—about deep enough to get out of the rain, and no more.

Early explorers had often commented that the St. Peter Sandstone looked like sugar—its white, rounded grains are uncemented and easily crumble off. While many caves are known from this sandstone, this one is unique in some respects. Everywhere else, the St. Peter Sandstone is white in color, suggestive of its purity, but at Sugar Cave it's filled with wild swirls of reddish iron pigment, known as Liesegang bands. The bands result from the oxidation and precipitation of iron from groundwater as the water approaches the surface of the rock exposure. Much of the iron comes from the mineral pyrite (iron sulfide) in the limestone layer above the sandstone.

Andrew Clemens, of McGregor, Iowa, mined sand for his well-known sand paintings, from Sand Cave, in Iowa. Clemens filled glass bottles with sand, grain by grain, building-up elaborate pictures, and sold them for modest prices. Each color was collected separately at the cave and placed in its own bag. No glue was used, the grains being held in place by surrounding grains. What makes his work even more amazing was the use of round-top drug jars that opened on the bottom, so that he had to create the images upside-down! His best known sand painting, on display at the State Historical Society in Des Moines, along with his tools, is a portrait of George Washington on horseback.

The art of sand painting was taken up in Minnehaha Park by people of far less skill than Clemens. Their art was widely sold in the 1880s, to the extent that considerable caves were hollowed out in the glen, owing to the sand extraction. The only remaining cave is the rosy-colored Sugar Cave at Godfrey's Point. The Hennepin History Museum in Minneapolis held an exhibition of Minnehaha sand art in 2018-2019.

MILL RUINS PARK
HENNEPIN COUNTY

During the half-century between 1880 and 1930, Minneapolis was the flour-milling capital of the world, drawing upon the hard spring wheat of Minnesota and the Dakotas and milling it into flour through the water power generated at St. Anthony Falls. The visitor to Mill Ruins Park in Minneapolis today, however, should realize that much of what he or she sees was invisible under a gravel yard for several decades, removed during creation of the park.

The first signage you come to, "Tailrace Skyline," near the parking lot, describes the early history of the milling area, presenting a "cutaway view" of a "waterpower canal and tailrace." This signage is crucial in understanding the basic terminology of milling and terms such as "tailrace." You learn that in 1857, the Minneapolis Mill Company began digging the First Street Canal parallel to the Mississippi River. Underneath it, the First Street Tunnel was dug. The canal was covered over with wooden planks where it ran along First Street (this plank road has been recreated just outside Mill Ruins Park). Water, diverted from the river, entered the upstream end of the canal through a gatehouse (which filtered out debris such as logs), was conveyed to the individual mills by branch canals, and then spilled down through shafts into the tailrace tunnels below, spinning turbines on the way down. The water then flowed through the various tailrace tunnels which merged to form the First Street Tunnel. The latter emptied into an open-air channel that carried the water back to the river, which now forms the axis of the park.

Farther along, up near the historic Stone Arch Bridge, built by railroad magnate James J. Hill in 1883, is the plaque labeled "Minneapolis Underground," which bears a detailed description and photos of the City Water Works tunnel—the big entrance right in front of you that is discharging large volumes of water. A former pumping station back in the early days of Minneapolis, before sewage had tainted the Father of Waters, drew drinking water from the river in the downtown area itself. The river became polluted, however, and in 1904, upon completion of the Columbia Heights plant farther upriver, this downtown waterworks was abandoned.

A map of the canals and tunnels is shown on this signage. After the tunnels were abandoned by the milling industry, but before they were gated off for Mill Ruins Park, it was easy for urban explorers to propel themselves up and down the flooded tunnels in a boat with a pole, like a subterranean gondolier, and hence the tunnel complex was dubbed "Subterranean Venice" by them. In the corner where the limestone ledge is visible, you can see the chief outlet for this "Venice," but because river water is no longer diverted through the turbines, only a trickle of water is seen flowing through the gate. It would have been different back in the heyday of milling!

Several gated tunnels at Mill Ruins Park mark the entrance to a veritable "Subterranean Venice." (Photo: Greg Brick)

Farther south along the walking loop a different system prevailed. There, each mill, although supplied by the First Street Canal, had its own tailrace tunnel that flowed independently out to the river, instead of joining a trunk tunnel. You can walk right up to a gated example of these solo tunnels and peer into the void.

As a bonus, you can see the great stormwater outfall for the city of Minneapolis, as you approach the parking lot on your way back. Look for the rounded arch at water-level in the canal near where it joins the river. This outfall became the jumping-off point for the wetsuit-clad sewer expeditions by urban explorers to Schieks Cave, the largest cave under downtown Minneapolis, 75 feet below street level. While not on any park signage, the story is told in Greg Brick's 2009 book *Subterranean Twin Cities*.

Directions:	Coming from downtown Minneapolis, take Portland Avenue toward the river. Drive downhill under the Stone Arch Bridge and take a right, heading downhill, past the U.S. Army Corps of Engineers Visitor Center. Drive all the way down to the small parking lot at the very end of the road. From there you can walk the easy horizontal loop through the park with its gated tunnel entrances.
Seasons/Hours:	Year-round, dawn to dusk.
Length:	Loop is one half mile long.
Precautions:	None.
Amenities:	No restrooms.
Information:	Minneapolis Park and Recreation Board, 2117 West River Rd., Minneapolis, MN 55411. Phone: 612-230-6400. Website: www.minneapolisparks.org.

THE LOST WORLD SAFARI

An archipelago of sewer caves exists in the sandstone below downtown Minneapolis. In 1939, *Minneapolis Journal* photographer David Dornberg went on a "Camera Safari," as he called it, through these murky caverns, now called Schieks Cave. He described it as "a 'lost world,' weird and spooky—the darkest spot for adventure into which my four years as a *Journal* cameraman ever led me."

Schieks Cave is the largest cave under downtown Minneapolis, extending for a city block through the St. Peter Sandstone, but there are other, smaller caves nearby. Carl J. Illstrup, city sewer engineer, who discovered the cave in 1904, described it as a "cave shaped like an inverted bowl," a description that seems puzzling to anyone who has actually been there. In 1907, journalist Jack Longnecker penned an atmospheric discovery narrative titled "In Caverns of Eternal Night." In 1931, another journalist, Robert Fitzsimmons, waxed poetical about "the beauties of the sewer system" and described Illstrup as "the ruler of this fantastic world." The discovery of the cave in 1904, during the excavation of the North Minneapolis Tunnel, when the crews braved "the lethal breath of deadly gases," is presented as the highpoint of Illstrup's life.

Reportedly, Schieks Cave was kept a secret for years because city officials feared the public would think downtown Minneapolis was built on a thin shell that would plunge into a hole in the earth. By 1921, it had been reported that "the entire business portion of the city is built over a series of subterranean lakes and caverns as mysterious and baffling as the Mammoth caves of Kentucky."

In 1983, the Minnesota Speleological Survey explored Schieks Cave and their account of the gloomy cave whose walls were black with cockroaches, with the roaring sewer that ran under it, only fired the imagination of later generations of urban explorers. A horrific journey through the sewers to Schieks Cave in the year 2000 was described in Greg Brick's 2009 book *Subterranean Twin Cities*. He reported pyramid-style concrete piers supporting the ceiling in many locations, an example of the vaguely Egyptian appearance of deep sewer architecture. Other features included the bubbling Black Sea and the cave's benighted tenants, the fly and worm ecosystem.

Schieks Cave is a "lost world" far below the streets of Minneapolis. (Photo from the Greg Brick Collection)

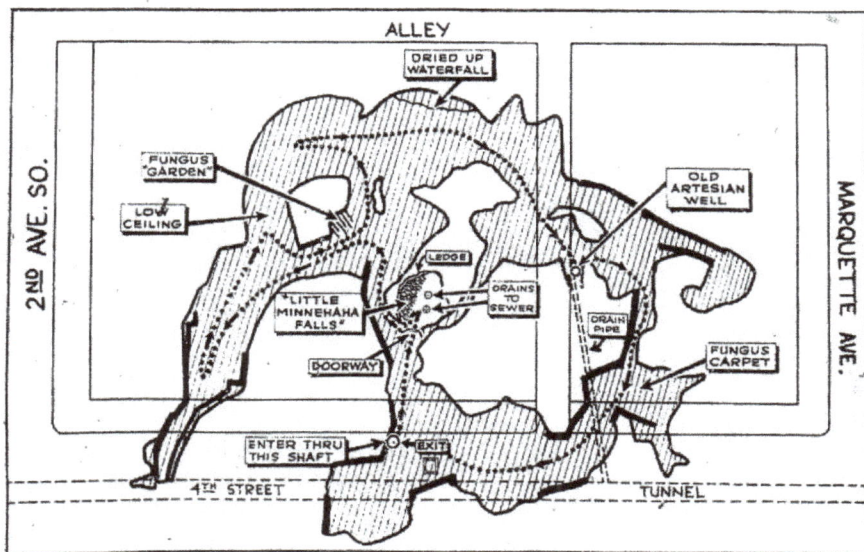

ROUTE OF CAMERA EXPLORATION UNDER MINNEAPOLIS LOOP

Dornberg's 1939 map of Schieks Cave.

CHUTE'S CAVE AND MILLRACES, PILLSBURY PARK
HENNEPIN COUNTY

While no part of the historic Chute's Cave is publicly visible, the best place to ponder the bizarre story is among the abandoned mill races at Pillsbury Park in Minneapolis. Descend the wooden stairways and bridges onto a small wooded island and look back toward the river banks, with the historic Pillsbury A Mill looming above you. As you visit this city park, look at the overgrown, spray-painted exits of the flour mill tailraces on one side, and those for the saw mills, on the other. Sadly diminished streams of water may be seen trickling out, a far cry from when the mills were running full blast—back when Minneapolis was proclaimed the Flour-Milling Capital of the World!

Minnesota caves made a dramatic debut on the national stage after the American Civil War. On December 10, 1866, a hoax article titled "Curious Discoveries in Minnesota," appeared in the *New York Herald*. Reuben Nesmith was digging a potato bin in the town of St. Anthony, now part of Minneapolis, when he struck an iron trapdoor, "beneath which a spiral stone staircase led down into the earth." Along with his brother-in-law, Luther Chamberlain, he descended the stairway of 123 steps and they "found themselves in a narrow horizontal passage, dug in the white sand." They entered "a

Pillsbury Park with its millraces was the scene of a national cave drama. The historic Pillsbury flour mill looms above. (Photo: Greg Brick)

spacious artificial cave, also excavated in this white sand," which became known as Nesmith's Cave. Successive chambers contained relics of a prehistoric civilization, including iron and copper implements, a colossal human figure, hieroglyphics, a stone sarcophagus and a sacrificial altar. When the sarcophagus was opened, a human skeleton was found, the bones of which crumbled to powder.

On January 9, 1867, the Minneapolis *Chronicle* ran a lurid embellishment of the Nesmith Cave hoax. In this version, Nesmith, Chamberlin, and the city council armed themselves with Roman candles and descended into the cave where they traversed the now familiar succession of chambers containing marvels. In one of them, "a huge stalagmite has been formed, we called it the tower of St. Anthony. It is a lofty mass 200 feet in circumference, surrounded from top to bottom with rings of fountain basins." The next chamber was even larger, enough to contain "the whole of our Catholic Church." A rocket was fired, exploding as it struck the immense dome, creating a shower of falling stars with "the roar of a cannonade."

Nesmith led the city council into ever stranger realms. He illuminated "a delicious little cave arched with snowy stalactites," in which there was a table "adorned with goblin knickknacks. It was the boudoir of some gnome or coquettish fairy." The next chamber was groined with gothic arches and paved with "globular stalagmites." "In a corner fountain," Chamberlin wrote, "we found the skeleton head and body of a serpent of uncreditable size." They passed into "another vaulted cathedral" that was flooded with "a strong iron water." Once again, Chamberlin displayed a gift for vivid imagery: "This dark lake lit up by the blaze of a dozen Roman candles, and reflecting the flashing walls of the cavern, would have made a picture for Barnum." A nearby skeleton, eight feet high, caused him to wax philosophical: "Whether he was a lost traveler, an absconding debtor, a suicidal lover, or a wretched murderer seeking concealment from vindictive pursuers, no one can tell."

Despite the hoax, Nesmith's Cave was based on a real cave that still exists today, Chute's Cave. Named after Dr. Chute, who supervised the excavation, it would become part of an amusement resort. In the Minneapolis *Tribune* for August 10, 1875, a "New Attraction" is advertised. "Mr. Mannasseh Pettengill has leased the famous Chalybeate Springs" and will "carry out a plan of improvements, which will make it more popular than it was before the war." As depicted on an engraving of the "St. Anthony Falls Mineral Springs," Pettengill's resort included a photographic gallery, observation tower, hotel, and bath. Warner's *History of Hennepin County* mentions "a fish pond and a few curiosities of the animal kingdom. The view of the falls with these extraordinary inducements, rewarded the tourist for the fatigue of descending the long stairway to the bed of the river, and the patronage of the swing, boat and restaurant compensated the enterprising owner."

In the Saint Paul and Minneapolis *Pioneer-Press* and *Tribune* for August 26, 1876, there is another advertisement for Chalybeate Springs. The attractions included "ice cream parlors and [a] cigar stand" and "a little building occupied by M. Nowack, as a photograph gallery, where may be found stereoscopic views of that locality." "The band plays at the springs every Saturday evening," it continues, "and with the grounds brilliantly illuminated, and the grand old Mississippi rolling and tumbling at your feet, the scene is a beautiful and impressive one." But now we read about "Chute's Cave—A Boat Ride of 2,000 Feet, Under Main Street." This was the first use of the name Chute's Cave. The advertisement continues:

> *If you have a desire to Explore the Bowels of the Earth, Mr. Pettengill can accommodate you in that particular also. The mouth of the "Chute Cave" is just below the springs, and the bottom of this cave is covered with about eighteen inches of water. For the moderate sum of ten cents you can take a seat in a boat, with a flaming torch at the bow, and with a trusty pilot sail up under Main street a distance of 2,000 feet, between walls of pure white sand-stone, and under a limestone arch which forms the roof. It is an inexpensive and decidedly interesting trip to take.*

By hosting tours, Chute's Cave became the first and only show cave in Minneapolis history.

The summer of 1880 appears to have been the last season for Pettengill's resort, however. We read that a "large force of men is at work on the tail race, which runs along below the Chalybeate springs for several hundred feet, making sad havoc of the fountain and other paraphernalia connected with that resort." City sewers were connected into the tailraces, adding unsavory sights and smells. Pettengill's obituary states that he closed his resort "at a great sacrifice" and opened a farm in Todd County in the fall of 1881. "The beauty of this place [Chalybeate Springs]," it reported, "was ruined by the improvements made for water power and railroad purposes."

It's those very "improvements" that you can see in Pillsbury Park today. Chute's Cave itself remains deep in the sewer system, dangerous and difficult to access.

Directions:	Pillsbury Park along Main Street SE. Go down the wooden stairway and boardwalk to river level, and look back at the tunnel entrances.
Seasons/Hours:	Dawn to dusk.
Length:	Quarter-mile walking loop.
Precautions:	Rough and uneven trails. In winter the stairways leading down into the park are roped off for safety, owing to ice accumulations on the stairs.
Amenities:	St. Anthony Main has many shops and restaurants.
Information:	Minneapolis Park and Recreation Board, 2117 West River Rd., Minneapolis, MN 55411; 612-230-6400, www.minneapolisparks.org.

A THIRD KIND OF CAVE IN THE WORLD

A popular classification divides caves into natural and artificial categories. However, voids often develop adjacent to artificial excavations (for example, sewers) in loosely-consolidated geologic formations. Neither natural nor artificial in origin, these "unintentional caves" are best classified as "anthropogenic."

As preliminary, we should clarify the meaning of the term "natural cave." Among the general public, unlined, artificial caves dug into bedrock will often be called "natural caves" by laypersons because the avowedly natural rock surface is visible. Whereas to most geologists, natural refers to the space itself, not the walls.

There are two classic examples of anthropogenic caves in the St. Peter Sandstone:

Schieks Cave, a maze cave in the St. Peter Sandstone underlying the city of Minneapolis, has been attributed by some geologists to erosion of the poorly cemented sandstone into the North Minneapolis Tunnel. This tunnel is now the chief sanitary sewer for the downtown area. The cave was discovered by tracing white sand carried by the sewer back to its source. A natural spring found gushing in the cave could easily be what flushed out the sand. The time available to erode a void of this size, which underlies a city block, can be determined from the chronology. Construction of the tunnel began in 1889, while the earliest known record of the cave is a 1904 survey.

The case of Chute's Cave, also in Minneapolis, is more complex. A natural sandstone cave was discovered during construction of a tunnel by the St. Anthony Falls Water Power Company in 1866. The cave was enlarged artificially as shown by pick-marks on the walls. The cave was sealed off from the Phoenix Mill tunnel, a nearby water power tunnel, in 1874. Water leaked around the bulkhead, however, enlarging the cave to the extent that it caused Main Street to collapse, on December 23, 1880. The cave thus appears to have all three components: natural, artificial, and anthropogenic.

MLAC BOOK CAVERNS, UNIVERSITY OF MINNESOTA
HENNEPIN COUNTY

The University of Minnesota realized that it was running out of space for books and that the state legislature was tired of periodic requests for bonding. It chose the option of carving out three gigantic caverns in the sandstone below the West Bank Campus. Still, the legislature approved only two of them. Excavation began in 1997 and the new facility was dedicated by former Minnesota state governor, Elmer L. Andersen, in 2000. The above-ground portion, named the Andersen Library in his honor, sits atop the caverns, which are 85 feet below the surface. Greg recalls attending the inaugural tour. But you, too, can visit the book caverns. Do not expect to see the exposed cavern walls, however. All interior surfaces are covered.

Of the two caverns, one holds the university's archival collections and the other holds 1.5 million little-used books in shelving 17.5 feet high. Of the latter, 60 percent of the space is allocated to the university's own books, and 40 percent to 20 other libraries across the state. Collections come in on trucks through the lower portal, which you can see along the West River Road. The great space-conserving secret is that the books are shelved by size rather than subject. Consequently, the library classification systems you might be familiar with, such as Dewey Decimal, do not apply here and you would be bewildered to find any particular title were it not for the special barcodes.

The MLAC caverns hold millions of books. (Photo: MLAC)

Caverns are usually quite humid and so a dehumidifying system is required to keep moisture below 50 percent for cellulose materials. Also, MLAC was built near a Superfund site, the former Minnegasco facility, which led to contaminated groundwater seeping into the caverns, but a special treatment system was set up to handle the problem.

MLAC is nearly at capacity and while there is more space available in the rocks, further funding is unlikely at present. So a "Super De-Duping" project was begun to weed out duplicates. Another plan under consideration is to make the facility part of a national, rather than state, network. While much of what you access nowadays is stored "in the cloud" anyway, rather than in the rocks, it's essential to keep hard copies somewhere. Just in case.

Directions:	Andersen Library, 222 22nd Ave. South, Minneapolis, MN 55455. You will need to use one of the public parking ramps identified on the university's website.
Seasons/Hours:	Public tours are offered following First Fridays presentations, held in Andersen Library the first Friday of each month (October through December and February through May) of the academic year. Free.
Length:	Hundreds of feet.
Precautions:	Hardhats are provided for the tours.
Amenities:	Campus dining nearby.
Information:	https://www.minitex.umn.edu/Storage/About/Facility.aspx.

CARVER'S CAVE/WAKAN TIPI, BRUCE VENTO NATURE SANCTUARY
RAMSEY COUNTY

Carver's Cave became the baptismal font of Minnesota caving upon its exploration by Jonathan Carver (1710–1780), who visited what he called the "Great Cave" in 1766 and again in 1767. It became the earliest Minnesota cave in the published literature when the first edition of Carver's bestselling *Travels Through the Interior Parts of North America* appeared in 1778. Others, decades later, renamed it Carver's Cave. To Native Americans, who knew of it long before, it was called Wakan Tipi, the Dwelling of the Great Spirit.

On November 14, 1766, Carver reported his exploration:

> *About thirty miles below the Falls of St. Anthony, at which I arrived the tenth day after I left Lake Pepin, is a remarkable cave of an amazing depth. The Indians term it Wakon-teebe, that is, the Dwelling of the Great Spirit. The entrance into it is about ten feet wide, the height of it five feet. The arch within is near fifteen feet high and about thirty feet broad. The bottom of it consists of fine clear sand. About twenty feet from the entrance begins a lake, the water of which is transparent, and extends to an unsearchable distance; for the darkness of the cave prevents all attempts to acquire*

a knowledge of it. I threw a small pebble towards the interior parts of it with my utmost strength: I could hear that it fell into the water, and notwithstanding it was of so small a size, it caused an astonishing and horrible noise that reverberated through all those gloomy regions. I found in this cave many Indian hieroglyphicks [sic], which appeared very ancient, for time had nearly covered them with moss, so that it was with difficulty I could trace them. They were cut in a rude manner upon the inside of the walls, which were composed of a stone so extremely soft that it might be easily penetrated with a knife: a stone everywhere to be found near the Mississippi.

Looking out the entrance of Carver's Cave on a summer day in 2015. (Photo: Greg Brick)

Carver's liminal auditory and visual experiences, as recorded in his journal; the fact that the cave was located below a Native American burial ground, marked with petroglyphs, and was the source of a spring of pure water is more than enough for this cave to hold a special significance for anyone.

Carver carved the arms of the king of England among the petroglyphs in the cave. He thereafter visited St. Anthony Falls and ascended the Minnesota River, spending the winter with the Dakota Indians in a bark house. The following spring, he returned to the cave and on May 1, 1767, attended "a grand council."

In 1867, the Minnesota Historical Society held a "Carver Centenary" at the cave, and local druggist Robert O. Sweeny drew the first ever depictions of the cave from several perspectives. The guests rode a boat across the subterranean lake, admiring the famed rattlesnake petroglyphs by torchlight. Other petroglyphs, sketched by the antiquarian Theodore Lewis in 1878, represented men, birds, fishes, turtles, lizards, and so forth. Many of them were destroyed when the cliff was carved back to accommodate the railroad switchyards in front of the cave.

The most dramatic reopening of Carver's Cave occurred in 1913. John H. Colwell, president of the Mounds Park Improvement Association, was appointed to its "exploration committee" and promptly set about relocating the cave. On November 5, 1913, Colwell reopened Carver's Cave to the rays of the setting sun. Colwell's goal was to commercialize the cave, stringing lights and building a flight of stairs down the bluffs from Short

Street. A gaudy electric sign on the bluffs would be visible from the downtown area, attracting even more visitors. But the plan never materialized.

A journalist, Charles T. Burnley, drafted a conjectural map of the alleged discoveries in 1913, and in his crude cartography, Carver's Cave resembled the stomach of a cow with its various chambers. The Burnley map would be the starting point for others many years later. Getting into those rooms—especially that elusive and spectacular waterfall room at the very back—became quite a draw for local explorers.

Carver's Cave was repeatedly lost to sight over the years owing to detritus sloughing down from the bluffs above. On September 16, 1977, the cave was again relocated and opened with a backhoe as part of an official city bicentennial project, and Native Americans visited it the next day. Double steel doors were erected, which in the coming decades were themselves buried by a debris fan of the sloughed detritus.

In 2016, the 250th anniversary of Carver's first visit to the cave, remotely controlled floating rovers with lights and video recorders were used to produce YouTube footage of the remote corners of the cave. Human explorers, whose feet stirred up the fine silt on the bottom of the subterranean lake—thus clouding the water—could not get this clarity of footage. Old graffiti could be read below the water line, and it was hoped that petroglyphs, too, would be found. The prospect of opening the flooded rooms at the back end of the cave made this into an "Oak Island" mystery—though without the fanfare associated with that History Channel program!

The subterranean lake in Carver's Cave is home to many organisms. The popula- tions of amphipods (also called scuds or freshwater shrimp), pigmented isopods, white flatworms, and various gastropods, seem to be supported by the leaf litter that blows in through the entrance and decays in the lake. The scuds partake of this detritus, or "junk food," in turn supporting other creatures. The cave also serves as an overwintering refuge for frogs and mosquitoes. Beavers have been observed inside the cave amassing stick caches. But the blind crayfish reported in 1913 have not been seen since.

In 2005, Carver's Cave became the centerpiece of the new Bruce Vento Nature Sanctuary, named for the congressman who represented the East Side of St. Paul for many years. Established on what had been a railroad yard, the invasive buckthorn was removed and the original prairie restored by countless hours of volunteering. And while Carver's Cave is commemorated by an historical marker atop Dayton's Bluff, with a spectacular view of the city, the cave is nowhere near this marker. Instead, hike through the sanctuary itself, where several other historic caves can be glimpsed along the one mile of trails.

The first cave you come to on your way to Carver's Cave is the **North Star Brewery Cave**, easily recognized by its black iron gate. Carved out as a lagering cave for the North Star Brewery, whose foundations can still be seen in the weeds, the caves were abandoned about 1900. The cave was carved around a natural spring which still flows, providing a walk-in swamp cooler for the storage of beer. Today, watercress grows abundantly in the spring run exiting the cave, and is harvested by the local community for use in salads. Nearby, a stone stairway leads to an abandoned limestone quarry immediately above the cave, which for many years had been a major hobo jungle, furnished with discarded sofas. This cave was occupied during the Great Depression by homeless men, and is frequently confused with Carver's Cave. It's not far from Westminster Junction, at the intersection of rail lines from Chicago, Duluth, and the West (See TUNNELS TO THE PRAIRIE). Its most notorious inhabitant was an FBI fugitive on the lam, who was captured in 1950. In 1992, a homeless man was found floating in the lake inside the cave, his dead body covered with devouring shrimp.

Proceeding farther along the gravel path, you'll come to the Sand Castle, a towering, strangely eroded part of the white sandstone bluffs. It began as a limestone quarry, but after this hard caprock had been removed, water began sculpting out towers. This is the collecting locality of paleontologist Fred Sardeson, one of the few who had the patience to find molds of fossil clams in the St. Peter Sandstone. Don't bother searching for them, as even trained paleontologists have been unable to duplicate his feat!

Finally, a bit farther on, you'll come to the last pond in a series—often covered with algae in summer and ducks in winter—and the famous **Carver's Cave**, described above. It's unmistakable because of the rusted steel doors (dating from 1977) in front of the cave, although they are sometimes obscured by vines. You may see red cloth tied in the bushes thereabouts, left by Native Americans as an offering to the Great Spirit.

Carver's Cave has been "open" and "closed" over the years, referring to the opening around the edges of the steel doors. It was open from 1977 to 2009, when it was sealed by the City, then found dug open again in 2015. If it's open when you visit, you can get a glimpse of the subterranean lake by crawling on all fours about one body length into the constricted cave entrance, where the spring water is gushing out around the left side of the steel doors. Expect to get sand in your hair and wet shoes.

As of this writing, funding has been secured for the construction of the long-awaited Wakan Tipi Visitor's Center at the Bruce Vento Nature Sanctuary, which will hopefully open in the next few years. Stay tuned!

Directions:	Park at the paved lot under the Kellogg Avenue Bridge in St. Paul, which you can access from I-94 at the Mounds Park exit, turning off onto Commercial Street, which loops downhill to its intersection with East Fourth St., where you'll see the sign.
Seasons/Hours:	Year-round, dawn to dusk.
Length:	One mile loop out to the cave and back.
Precautions:	On a hot summer day, walking to the cave across the prairie can cause heat stress, bring bottled water.
Amenities:	Portapotty at the parking lot. New interpretive center soon.
Information:	St. Paul Parks & Recreation Department, City Hall Annex, 25 West 4th St., Suite 400, St. Paul, MN 55102. Phone: 651-266-6400. Website: www.stpaul.gov/departments/parks-recreation.

FOUNTAIN CAVE HISTORICAL MARKER
RAMSEY COUNTY

Fountain Cave was sealed in 1960 when the highway department constructed the present Shepard Road over the top of it. The Minnesota Historical Society erected the historical marker for Fountain Cave over the buried ravine in 1963. A good place to contemplate the early history of the city!

On July 16, 1817, on his way up the Mississippi River from Prairie du Chien to reconnoiter the Falls of St. Anthony, and again the following day on his way back down, Major Stephen H. Long (1784–1864), of the newly created U.S. Corps of Topographical Engineers, disembarked from his "six-oared skiff" in what is now St. Paul to explore something mysterious that perhaps the local Dakota bands had told him about.

A few miles below the confluence of the St. Peter's and Mississippi Rivers there was a gap in the bluffs where a small stream, later dubbed Fountain Creek, met the great Father of Waters. Disembarking, Long and his men ascended the ravine for more than one hundred yards. They came to a natural amphitheater of snow-white sandstone whose walls, 40 feet high, formed three-fourths of a circle, making it seem as though they were standing at the bottom of a gigantic pit. Swallows darted from innumerable holes in the cliffs. The creek issued from a Gothic cave entrance 16 feet high and about as many wide.

Fountain Cave in 1850, the earliest cave image in Minnesota. (Courtesy of Minnesota Historical Society)

They passed through the cave's pearly-white gates and entered a large winding hall about 150 feet long. The sharp drop in temperature came as a welcome relief on this hot summer day. At the far end of the room, they crawled through a narrow passage that opened into "a most beautiful circular room" about 50 feet in diameter, where their candles flickered against the walls.

"The lonesome dark retreat," Long later wrote in his journal, was cheered by the "enlivening murmurs" of the "chrystal [sic] stream." Wading in icy cold water up to their knees, the soldiers continued along the meandering passage, encountering more rooms of a circular form and penetrating about 200 yards before their candles went out. They halted, and began to grope their way back in stygian darkness. The U.S. Army, in the persons of Major Long and his men, had just discovered what was thereupon named the Fountain Cave.

Many others followed over the years. By the time the famous Pierre "Pig's Eye" Parrant—depicted with his pirate-style eye patch on countless beer cans in our own day—arrived on the scene, Fountain Cave already had a respectable bibliography. The 1837 treaty with the Ojibway having opened for settlement the triangle of land between the St. Croix and Mississippi Rivers, Parrant staked a claim at this cave because it was the nearest point to Fort Snelling that was not actually on the military reservation, thus shortening the distance for the soldiers to whom he sold whiskey.

Parrant was a French Canadian voyageur who attempted sedentary habits in his old age, but he did not actually live in Fountain Cave. On the contrary, much of his supposed historical importance rests in the fact that he erected a log cabin, one of the first buildings on the site of what is now St. Paul, on or about June 1, 1838. Often loosely described as a "saloon," it was sited at the mouth of the secluded gorge so that potential customers could see it from the river. Some squatters at Camp Coldwater, near Fort Snelling, soon moved downriver to join Parrant, and cabins began to sprout like mushrooms at the cave. But since the platting of the city of St. Paul actually began in 1849 with "St. Paul Proper," in what is now the downtown area, and not at Fountain Cave, the traditional claim that Parrant founded the city is dubious.

Strange things were reported of Fountain Cave about this time, providing the first Minnesota ghost stories. "In later years," a 1920 newspaper clipping asserted, "children of the settlers playing within its chambers heard shrieks of the dying Indians, just as they had occurred hundreds of years before, when put to death by their enemies and saw white-robed spectres floating from chamber to chamber, it's said. Even now, after one has found his way down the tortuous sides of the river bank to the spot where a few fishermen's cottages still stand, children of the neighborhood will tell of the strange happenings that go on in the ravine at the mouth of the cave."

The years from 1850 to 1880 were Fountain Cave's golden age. It became a fashionable Victorian cave—the first commercial or show cave in the Upper Midwest.

Minnesota governor Alexander Ramsey himself went spelunking there, as related in Elizabeth Ellet's *Summer Rambles in the West*, published in 1853. "A rustic pavilion stands in the woods," she wrote, "where lights can be procured to enter the cave." A footbridge over the ravine had been constructed. She compared Fountain Cave, which she called "Spring Cave," to "a marble temple" and its stream to "a shower of diamonds." Frederika Bremer, in her *Homes of the New World* (1853), described Fountain Cave as "a subterranean cavern with many passages and halls, similar probably to the celebrated Mammoth Cave of Kentucky. Many such subterranean palaces are said to be found in Minnesota." A letter to the *Congregationalist* of Boston for September 19, 1856, described a visit to the cave, mentioning "the torch of birch-bark which your guide manufactures for the occasion."

The oldest known graphic depiction of a Minnesota cave is a pencil and watercolor of Fountain Cave by an unknown artist about 1850. The most elaborate literary depiction of Fountain Cave was presented by the Galena, Illinois, journalist E.S. Seymour in his *Sketches of Minnesota, the New England of the West*, published in 1850. Seymour's description establishes the cave as an unbranched tube, wholly in the sandstone layer. Apart from widenings of this passage, called rooms, much of the passage was crawlway. There were four rooms successively decreasing in size upstream, of which he gave the dimensions. The third room back was the only named feature in the cave, called "Cascade Parlor" because it contained a waterfall two feet high; he suggested planking over the stream here to make it more accessible to visitors. He did not go beyond the fourth room, having penetrated an estimated distance of 60 rods (990 feet), but stated that he could hear a second waterfall in the distance.

In 1880, the newly formed Chicago, St. Paul, Minneapolis and Omaha Railroad—called the Omaha, for short—began building a roundhouse and repair shops in the triangle of land bounded by Randolph Avenue, Drake Street, and the river. The oldest and only complete map of Fountain Cave known to exist shows this facility already in place. Judging from this map, Fountain Cave is the longest (but not the largest, in terms of volume) natural sandstone cave in Minnesota, about 1,100 feet. The railroad mapped out the underlying cave in order to use it as a sewer for the overlying shops. Fountain Cave abruptly dropped off the fashionable itinerary!

Fountain Cave is a natural cave formed by groundwater washing away sand grains, something that even the earliest observers recognized. But it was not until 1932 that St. Paul landscape architect George L. Nason added the capstone to our understanding when he described how the ravine at the cave's entrance—"the beautiful little valley," as he lovingly called it—was "formed by the caving in of the roof at various times."

While Fountain Cave remains sealed, desperate attempts were made to access the cave through the sewers, a "wild goose chase" described in Greg Brick's 2009 book *Subterranean Twin Cities*.

Directions:	The cave is sealed but the historical marker is on the east side Shepard Road one-third mile south of its intersection with Randolph Avenue. As the former highway pull-out has been eliminated, the nearest place to park is along Erie Street at Randolph, or at Butternut and Sumac streets.
Seasons/Hours:	Year-round.
Length:	1,100 feet, but the entrance is sealed.
Precautions:	Highway traffic; there is no convenient vehicular pull-out so the marker can only be reached by walking.
Amenities:	N/A
Information:	Minnesota History Center, 345 Kellogg Blvd. West, St. Paul, MN 55102. Phone: 651-259-3000. Website: www.mnhs.org.

The Giant Beaver Cave wasn't giant, beaver, or cave, but potentially a supremely important archeological site. (Courtesy of the Science Museum of Minnesota)

THE ALMOST-FAMOUS CAVE

The Giant Beaver Cave in the Highland Park neighborhood of St. Paul would have easily ranked as the oldest and most interesting archeological and bone cave in the Midwest if its earlier prospects had panned out. But it wasn't really a cave, and it wasn't a true beaver that lived there. Nor did Ice Age man live there, as some alleged.

On July 15, 1938, Works Progress Administration (WPA) laborers encountered the site while constructing a road into Hidden Falls Park. Dr. Louis Powell, curator at the St. Paul Institute, correctly identified the bones as those of the extinct giant beaver. This beaver occupied a wide swath of North America after the last Ice Age and was about the size of a black bear, its generous dimensions in line with the other megafauna of the day. But some of its habits were not those of the living species of beaver we are familiar with. And that news was topped when Clinton Stauffer, a geology professor at the University of Minnesota, announced that a "sharp bone splinter" found at the site could be a sign of man from 20,000 years ago. If that were the case, this would rank among the oldest archeological sites in the Americas!

But that was before radiocarbon dating was developed as a way of telling how old something was. Bruce Erickson, a paleontologist from the Science Museum of Minnesota, restudied the skeleton and in 1967 reported an age of 10,320 years for it. The so-called cave was really only the Platteville Limestone waterfall ledge of Hidden Falls, which apparently had collapsed over the beaver, trapping and preserving its remains, which are now mounted at the museum. And while there's no direct association with the Paleo-Indians—the tenuous splinter notwithstanding—the beaver lived at a time when they are known to have been in the region.

Owing to further "improvements" since then, nothing remains of the cave or the site in Hidden Falls Regional Park.

IRON MINE, MINNESOTA HISTORY CENTER
RAMSEY COUNTY

This re-created iron mine is one of the best and most accessible "underground" things to do in downtown St. Paul. On the second floor of the Minnesota History Center, as you enter the G. & M. Wells Gallery, the entrance to the iron mine is on your right, filling one corner of the gallery with its zigzag passageways. Grab a hardhat from the rack (not strictly necessary, it's merely for atmosphere!) and enter the dark mine. The path is wide and level and easy for everyone.

Minnesota led the world in iron production during World War II. There are three important Minnesota iron ranges: Cuyuna, Mesabi, and Vermilion, plus the Gunflint Range which however saw little mining activity. The fact that this recreated mine is meant to be underground suggests that it's likely on the Vermilion Range (see SOUDAN MINE) because on the Mesabi Range most of the ore was extracted in open pits, not underground workings. The walls are painted to represent banded iron formation, with alternating chert and hematite that was laid down in a worldwide "iron age" two billion years ago during the Precambrian. It was no coincidence, as Earth's atmosphere began to acquire oxygen about this time from plants having evolved, so iron precipitated from seawater everywhere about the same time. This contrasts strongly with how the much younger sinkhole iron ores formed in southeastern Minnesota (see GOETHITE WMA).

This iron mine self-paced tour is organized around roleplaying each successive profession that worked in the mine. First is the geologist, telling you what blue vein to pursue. Next you play the driller, who drills holes in the wall for dynamite sticks, so the ore can be blasted free. Next (our favorite!) the blaster, with awesome sound effects as you push the plunger down after having loaded the six drill holes with imitation dynamite sticks. Then comes the backman, who taps the roof with a pole after the blast, to knock down loose rock, and finally the mucking cars, seeming to stretch to infinity owing to the effect created by mirrors. Signage reminds you that the average miner made two dollars for a 10-hour workday, laboring in candlelight and working with mules.

Directions:	Minnesota History Center, 345 Kellogg Blvd. West, St. Paul, MN
Seasons/Hours:	Year round, call or visit website for latest hours. Free, but other exhibits in the museum may charge for admittance.
Length:	100 feet.
Precautions:	A hardhat is provided but not really necessary
Amenities:	Restaurant, museum shops.
Information:	Minnesota History Center, 345 Kellogg Blvd. West, St. Paul, MN 55102. Phone: 651-259-3000. Website: www.mnhs.org.

Visitors can walk through this iron mine replica in downtown St. Paul. (Photo: Greg Brick)

Mirrors create the illusion of an endless line of ore cars at the Minnesota History Center. (Photo: Greg Brick)

SANITARY SEWER, SCIENCE MUSEUM OF MINNESOTA
RAMSEY COUNTY

Two hundred feet below the streets of St. Paul there's a huge sanitary interceptor. Before we go any further, however, a short digression on sewer jargon is required. There are two kinds of sewers, sanitary and storm. The term "sanitary" is a euphemism because it refers to raw sewage, which most of us would of course consider "unsanitary." The sewer is "sanitary" in the sense that it's connected to a sewage treatment plant, not discharging waste into nearby water bodies and polluting them. And the sewage is thus said to have been "intercepted," hence interceptor. On the other hand there's the storm sewer, often called a storm drain, which carries rainwater to water bodies by means of an exit, called an "outfall." In the bad old days, most sewers did double duty, leaving water bodies severely polluted, but in the Twin Cities, the separate system was introduced when the municipal treatment plant at Pigs Eye, near St. Paul became operational in 1938. It treated the sewage of both Minneapolis and St. Paul. Occasionally, however, the two kinds of sewer would still overflow into one another, so to further reduce river pollution an on-going effort has been made in recent decades to separate them even more thoroughly.

You can glimpse a short re-creation of this tunnel at the museum. On display are a variety of objects that get into the sewers and end up being filtered out when they reach the treatment plant. Items include buoyant objects, usually toys. Some of the items are so large they could not have been "flushed" so it's a minor mystery how they entered the system in the first place.

Directions:	120 Kellogg Blvd. West, St. Paul, MN.
Seasons/Hours:	Call or visit website for the latest hours. Fee.
Length:	Several feet.
Precautions:	N/A
Amenities:	Omnitheater, café, and gift shop.
Information:	Science Museum of Minnesota, 120 Kellogg Blvd. West, St. Paul, MN 55102. Phone: 651-221-9444. Website: www.smm.org.

THE SANDHOG'S PARADISE

The 1930s were the golden age of sewer caves, considering the natural sandstone caves discovered during the course of the big sanitary sewer projects, begun both to clean up the Mississippi River and help get us out of the Great Depression. In 1935, for example, the sandhogs, as tunnel laborers were then called, struck what remains the largest known cave under Minneapolis. Although the name Channel Rock Cavern was given to it years ago, this name has no obvious historical roots. Many cavers to this day simply call it the 34th Street Cave, as it's located near East 34th Street and West River Road. To the sandhogs, the cave was a blessing in disguise because it gave them a convenient place to dump their sand as they continued digging the tunnel. In pre-settlement times, the cave had obviously been open to the riverbanks and then had slumped shut with material sloughing from the slopes above.

Physically, Channel Rock Cavern is a great L-shaped hall, 800 feet long, 20 feet high and 50 feet wide. Sufficiently large to be considered for conversion into an underground city park. A new access point for the cave was created, a 52-foot shaft down from West River Road, capped with a hexagonal lid,

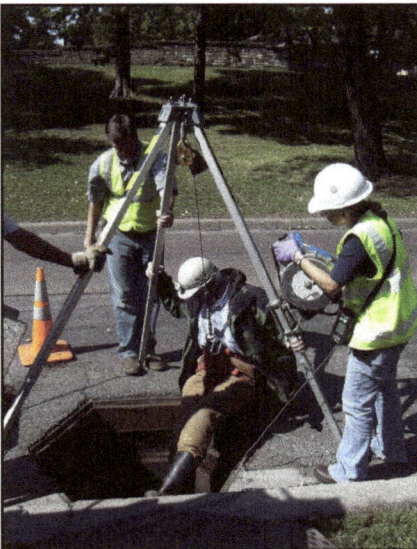

A university professor is lowered into Channel Rock Cavern, 2008. (Photo: Greg Brick)

The shaft into Channel Rock Cavern. (Photo: Greg Brick)

THE SANDHOG'S PARADISE (CONTINUED)

and just under it, a heavy slab of concrete that required a winch to remove. The cave was only rarely visited after that, so any visit became newsworthy, as in 1972 when sewer workers discovered "several fake graves" that someone had set up inside the cave, probably in the 1930s. In 1979, WCCO broadcasted the first televised visit to the cave. Greg was able to visit this cave by hiking through the sewers in 2007, another horrific experience described in his 2009 book *Subterranean Twin Cities*.

Investigating the origin of Channel Rock Cavern. (Photo: Greg Brick)

BATTLE CREEK REGIONAL PARK
RAMSEY COUNTY

In 1842 there was a locally famous Indian battle on the outskirts of the frontier settlement of St. Paul. This clash between rival tribes of Indians gave its name to the sprawling Battle Creek Regional Preserve, covering nearly three square miles and crisscrossed by hiking trails through a variety of natural habitats.

At the lower picnic grounds on Point Douglas Road you'll see the dazzling white sandstone outcrops like petrified snow. Every visitor over the years seems to have recorded their names and sentiments on those outcrops, a sort of stony newspaper for the curious to read. Informal but well-used trails run along the side of the bluffs on both sides of Battle Creek. Follow those trails and you'll soon come across several artificial caves that were dug perhaps as root cellars. One of them was short and swarming with crane flies, which look like giant mosquitoes but are quite harmless. Another is 30 feet long, roomy with bedrock benches. During a visit on a hot day in 2019 whole families were seen touring the cool caves.

Battle Creek itself begins near the 3M Global Headquarters in the city of Maplewood, miles away. Flowing down through the rugged ravine before reaching the cave grounds, it vanishes into a grate and passes through a culvert under US

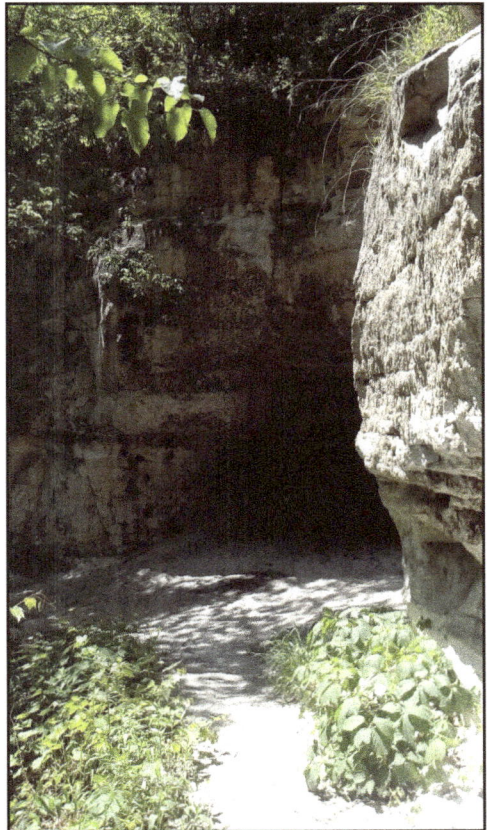

The peaceful caves of Battle Creek were dug in the snowy white outcrops sometime after the Indian wars. (Photo: Greg Brick)

Highway 61, finally entering Pig's Eye Lake on the floodplain of the Mississippi River. The lake was named after the very same "pig-eyed" voyageur we met in the Fountain Cave entry of this book!

Directions:	From US Highway 61 south of St. Paul, turn east at Lower Afton Road and then immediately north on Point Douglas Road, which ends in a loop at the lower entrance to the park. Other park entrances involve lengthy hikes to get to the caves.
Seasons/Hours:	Year-round, dawn to dusk.
Length:	From several feet to several tens of feet in length.
Precautions:	Obey all park policies and signage. Avoid caves that are signed for non-entry. You can still look in through the cave entrance even if you can't enter a cave.
Amenities:	Picnic tables, hiking, mountain biking.
Information:	Ramsey County Parks & Recreation Department, 2015 Van Dyke St., Maplewood, MN 55109. Website: https://www.ramseycounty.us/residents/parks-recreation/parks-trails/find-park/battle-creek-regional-park

SWEDE HOLLOW PARK
RAMSEY COUNTY

Hiking the paved trail looping through this historic, wooded ravine, with its shear walls, you begin to wonder what natural process could have led to its formation. Much of the geological evidence has been altered by human activities so it's not easy to tell. Was it carved by a stream? Is it a waterfall retreat gorge? In fact, you might be walking through the remnants of a gigantic collapsed cave!

But the ravine is rich in historical associations as well. Even before the American Revolution, the Swedish Academy of Science had proposed the prize question, "What can be the causes of such a multitude of Swedes emigrating each year?" One cause was certainly the famines that decimated the Swedish countryside in the wake of the Little Ice Age, which caused crop failures. The State of Pennsylvania was once called "New Sweden" owing to Swedish emigration. St. Paul's Swede Hollow originated from a later wave of emigration.

Swede Hollow Park is now a peaceful, quiet, shaded sanctuary in the heart of the bustling city but it was once a bustling immigrant community. Scandinavians began arriving in the 1850s and the deep ravine protected them from the worst blasts of winter and kept them cool in summer. The Hollow became a focal point for subsequent immigrant groups as well, including Irish, Italians, Poles, and finally Mexicans. Many of them worked for the St. Paul & Duluth Railroad (the "Skally Line") whose tracks ran

Looking through the historic helical stonework arches into Swede Hollow—a giant collapsed cave?
(Photo: Greg Brick)

through the ravine. Once they learned English they usually left the Hollow for other neighborhoods.

Shanties and outhouses lined the banks of Phalen Creek, which ran through the Hollow, until it was decided that since the community lacked infrastructure such as water and sewer service, it was a health risk. In 1956, the St. Paul Fire Department torched the Hollow in a mighty conflagration. There was little left to see afterwards except for the entrances to the abandoned Drewry Caves, formerly used for storing ale and porter. Even those were sealed in 1964, after some tragic cave deaths. The only trace of the caves today is a patch of whitish sandstone visible along the hiking trail. Standing mute witness to the former community is the assortment of catalpa, Kentucky coffee, and other urban trees planted by the immigrant groups.

Residents of Swede Hollow, lacking water service, and not being able to use the heavily polluted stream, used a spring for drinking water. Nowadays, you can see these springs off to the side of the trail where the ravine takes a big elbow bend about half way through the park. The spring water, emerging from the sandstone, collects in pools thick with watercress, enough to form floating bog mats, before draining under the trail through a culvert, and joining the stream. A new wave of immigrants from the 1980s onwards, Hmong from Laos, though not living in the Hollow, frequently harvest the watercress that grows here, collecting it in plastic shopping bags. Technically an invasive plant, watercress makes a tasty addition to salads.

At the south end of the Hollow you'll find the double-barrel "7th Street Improvement Arches," a National Historic Civil Engineering Landmark dating to 1884. Notice how the stonework is arranged in a spiral. At the other end of the ravine, where Minnehaha Avenue passes overhead, you'll see the tall stack of what had been Hamm's Brewery and is now Flat Earth Brewery. Unless you're hiking the regional trail, there's not much to see beyond this point.

Directions:	The park has multiple entrances but the parking lot at the south end is easiest to find, at the intersection of East Seventh Street and Payne Avenue. Park there and follow the trail signs. Passing through the double-barrel arches, you're now in Swede Hollow. If you start at the north end, turn off of Payne Avenue at Preble, then to Beaumont where it intersects with Drewry Lane, park, and walk down the sloping pedestrian underpass (called the Drewry Lane Tunnel) into the park.
Seasons/Hours:	Year-round, dawn to dusk.
Length:	The park is a half mile long and its well-shaded, paved hiking loop is about one mile long.
Precautions:	Some homeless encampments in the woods; may not be safe after dark.
Amenities:	Interpretive signage at the parking lot. Park benches but no picnic tables.
Information:	St. Paul Parks & Recreation Department, City Hall Annex, 25 West 4th St., Suite 400, St. Paul, MN 55102. Phone: 651-266-6400. Website: www.stpaul.gov/departments/parks-recreation.

TROUT BROOK NATURE SANCTUARY
RAMSEY COUNTY

Trout Brook was once a surface stream that flowed several miles from McCarron's Lake in Roseville to the Mississippi River at St. Paul. Back in pioneer days, Edmund Rice built a mansion, called "Trout Brook," nearby, hence its name. Back then, the stream was not only good enough to support trout, it was good enough to drink. But now it's St. Paul's great subterranean river. The trout have been supplanted by carp swimming through the great storm drain!

An initial caveat is in order. The stream that you'll see flowing through the Trout Brook Nature Sanctuary is not the historical Trout Brook but rather stormwater from the surrounding neighborhoods, filtered through special iron-enhanced sand beds to remove nutrients that would otherwise escape to the Mississippi River and cause problems downstream, in the Gulf of Mexico's infamous Dead Zones. The treatment system and sanctuary, replacing the railyards that once marred the view, is a collaboration of the

Capitol Region Watershed District and many others. The actual stream tunnel itself is deep underground running along Interstate 35E, which borders the sanctuary on the east.

But we should back up a bit. Surface streams get buried and "lost" for a variety of reasons. Sometimes the land on which the stream flows is needed for other purposes. Or sometimes, as in the case of Trout Brook, the streams were not buried per se, so much as that the adjacent street grade just grew upwards around them over the years.

In a very real sense, of course, the former surface streams are not "lost" since they are still flowing as lustily as ever. Indeed it would take a very expensive feat of engineering to get rid of them completely. To truly eliminate a stream you'd have to fill the drainage basin itself, eliminating the low spot.

The immense amount of geological work that the Trout Brook precursor stream did in carving this low spot is best visualized in downtown St. Paul, several miles away. One of the most salient topographic features of downtown St. Paul is the mile-wide gap in the white crescent of sandstone cliffs along the Mississippi River. City Hall stands on a full thickness of bedrock, but the sandstone thins out where Kellogg Boulevard goes downhill, finally to vanish from sight altogether before reappearing in all its glory at Dayton's Bluff. Lowertown occupies the gap. But what created this gap in the first place?

Geologists long ago surmised this gap was carved by a pre- or interglacial precursor of the Mississippi River, flowing down from the north. The Mississippi has changed course several times in the past million years or so and has only lately carved its present gorge. The topographic depression left by its precursor, partially refilled with glacial sediments in the interim, became the focus of postglacial drainage, and the stream that now runs through the gap is called Trout Brook.

The Trout Brook valley was a blessing for the railroads, providing an easy gradient on which to lay tracks (see TUNNELS TO THE PRAIRIE). Railroads have so dominated this valley ever since that the land between Trout Brook and its neighboring Phalen Creek (see SWEDE HOLLOW PARK) came to be known as "Railroad Island." In 1893, city engineer George Wilson undertook the arduous task of formally burying the two streams, a task completed by others decades later. A short length of the original surface Trout Brook, before it vanishes into its tunnel, may be seen flowing out of McCarron's Lake near the St. Paul Regional Water Treatment Plant on Rice Street.

Directions:	1200 Jackson St., at Maryland.
Seasons/Hours:	Year-round, dawn to dusk.
Length:	The gravel hiking trail is a loop more than one mile in length.
Precautions:	On a hot summer day, walking across the prairie can cause heat stress, bring water.
Amenities:	Restrooms and interpretive signage at the parking lot.
Information:	St. Paul Parks & Recreation Department, City Hall Annex, 25 West 4th St., Suite 400, St. Paul, MN 55102. Phone: 651-266-6400. Website: www.stpaul.gov/departments/parks-recreation.

TUNNELS TO THE PRAIRIE
RAMSEY COUNTY

Some cities are renowned for their train tunnels, a prominent example being New York City, which spawned the so-called "Mole People," about which whole books have been written. This category of subterranean space is much less well represented in Minnesota but there's one place in particular that positively invites tourists for a peek, and that's the Westminster Junction viewing platform with its informative signage.

St. Paul was a great railroad hub at one time, and its Lowertown is still graced with large old warehouses from the days of the railroad barons, often now repurposed as artists' lofts. Here was the meeting point of trains from Chicago, Duluth, and the West. Where the lines crossed, near Westminster Street, some ran under others in tunnels. According to the signage at the overlook, four tunnels were built at this transportation bottleneck from 1862 to 1909, the longest of them 1,048 feet long. One of them was later filled in. Beyond was the upland prairie leading to Minneapolis.

At the viewing platform, one of the four panels, "Tunnels to the Prairie," describes why railroad tunnels were needed here, owing to the pinch point in the valley. The other

Railroad tunnels are visible from bridges over the Trout Brook valley. (Photo: Greg Brick)

panels describe life on "Railroad Island," the triangle of land where Trout Brook and Phalen Creek came together, and other topics.

From the platform on the Phalen Boulevard Bridge, you can look down upon a railroad arch built with the cream-colored Kasota Stone, dated 1888. The next bridge to the south, at Lafayette Road, will enable you to glimpse the entrance to a much longer tunnel. However, there's no viewing platform or signage here, and the nearest parking is along Otsego Street, half a block away.

If you have further interest in local railroad history, the Jackson Street roundhouse and shops, now the Minnesota Transportation Museum, is nearby, at Jackson and Pennsylvania avenues.

Directions:	The Westminster Junction viewing platform is located on the Phalen Boulevard Bridge, however, you'll not be able to stop. Nearest parking is along Olive Street, with a several minute walk uphill to the platform.
Seasons/Hours:	Year-round.
Length:	Tunnels are about one city block in length, but the ones visible from the platform are much shorter.
Precautions:	Do not attempt to walk through tunnels as you could be cited for trespassing. Also note that the Trout Brook valley is a major camping ground for homeless people, who have set up tents in wooded areas.
Amenities:	N/A
Information:	Minnesota History Center, 345 Kellogg Blvd. West, St. Paul, MN 55102. Phone: 651-259-3000. Website: www.mnhs.org.

THE SELBY STREETCAR TUNNEL

In 1890, the St. Paul Street Railway Company, with its plant situated near the Hill Street Station, began to electrify its lines, running the cables through yet another set of sandrock tunnels, supported on heavy iron frames. These tunnels were abandoned in 1953, when buses replaced streetcars. One sandrock tunnel carried cables all the way to another tunnel, the historic Selby streetcar tunnel, 1,500 feet long, which opened in 1907. The Selby tunnel allowed St. Paul's cable cars to get to the top of the plateau behind the downtown. After the demise of streetcars, the abandoned Selby tunnel was used by the homeless. You can see the now-sealed entrance to the Selby streetcar tunnel on the slopes below the St. Paul Cathedral. Occasionally, you can peek through holes in the wall made by curious visitors and get a glimpse of the vaulted interior of the tunnel. But please don't try making any holes yourself!

The abandoned Selby Streetcar Tunnel once sheltered a homeless community. (Photo: Tony Andrea)

MUSHROOM VALLEY, CHEROKEE AND LILYDALE REGIONAL PARKS
RAMSEY COUNTY

The city of St. Paul—much more so than its twin, Minneapolis—is known for its artificial sandstone caves, and the densest cluster of them is surely Mushroom Valley, across the Mississippi River from downtown St. Paul.

Mushroom Valley, according to the boast, was the largest mushroom-growing center west of Pennsylvania, or alternatively, west of Chicago. Sometimes it was called the mushroom capital of the Midwest. The mushrooms were grown in the more than 50 sandstone caves that punctuated the bluffs. Although called caves, they were artificial. Often begun as silica mines, these caves were subsequently used for mushroom growing and other purposes. According to newspaper columnist Oliver Towne (Gareth Hiebert), "A whole economy and countless legends lie locked from view inside those rustic cliffs." He dubbed them the "Ivory Cliffs" owing to the snowy whiteness of the St. Peter Sandstone.

Used in the broadest sense, Mushroom Valley is divided into three distinct segments: Plato Boulevard, Water Street, and Joy Street. Each segment has its own distinct flavor. The Plato segment, incorporating what had been the cave-riddled Channel Street before a 1970 replatting reset the street grid, is capped by Prospect Terrace with its historic houses and magnificent views of the city. The Water Street segment, running right along the river and under the High Bridge, had by far the largest caves, forming a labyrinth extending under what is now Cherokee Regional Park. The Joy Street segment, now vacated, can still be seen where an unmarked dirt road runs through the woods in Lilydale Regional Park. While most of these caves are very short—root cellars and such—Joy Street is anchored by large caves: Mystic Caverns at its eastern end and Echo Cave at its western end. Known better for its namesake acoustic effects, Echo Cave's passages served as brick drying tunnels for the St. Paul Brick Company and were gated as a bat hibernaculum by the DNR in 1989. Beyond that are the brick company's clay pits in the Decorah Shale, a bizarre industrial landscape but the city's premier ice-climbing venue because groundwater seeping from the cliffs freezes to form gigantic ice formations in winter.

The Mushroom Valley caves had been considered for bomb shelters during World War II, even before the Pearl Harbor attack. During the Cuban Missile Crisis of the early 1960s they were surveyed for suitability as nuclear fallout shelters, producing the only maps that we have today. The typical cave is a straight, horizontal passage about 150 feet long, often connected by cross-cuts to similar caves on either side, creating network mazes with multiple entrances. A cave operated by the Becker Sand & Mushroom Company was the largest of all, with 35-foot ceilings and nearly one mile of passages, its wonderful hybrid name capturing the chief dual usage seen throughout the valley.

An abandoned Joy Street root cellar seen in winter. (Photo: Greg Brick)

View from the Joy Street root cellar on a snowy day. (Photo: Greg Brick)

Not all the former sand mines were used for mushroom growing. Examination of city directories, insurance atlases, and real estate plats allowed a fuller picture of the diversity of people and businesses that inhabited Mushroom Valley. These sources reveal what each of the caves was used for, as it's fairly easy to correlate each street address with a particular cave entrance. The chief uses of the Mushroom Valley caves were mushroom gardening, cheese ripening, brewing, and as places of entertainment, especially nightclubs.

The hopeful comparison made by boosters in the post-mushroom era has been with Kansas City, Missouri, where a subterranean industrial park of several thousand acres, the so called "SubTropolis," was created from a room-and-pillar mine in limestone. Although this use of underground space is periodically suggested for Mushroom Valley, as it was most notably by the Condor Corporation in the 1980s, inevitably a counter-movement develops, arguing that the neighborhood will be undermined and collapse into the ground.

Directions:	This is only a drive-by of the sealed cave entrances. Drive the length of Water Street in Cherokee Regional Park, looking up into the woods.
Seasons/Hours:	It's easiest to spot any exposed cave entrances in winter, when not concealed by vegetation.
Length:	Various, but typically 150 feet.
Precautions:	The cave entrances are regularly sealed by the city and dug open again by locals. The valley is regularly patrolled and if you're seen entering a cave you may be given a citation by law enforcement. The fossil grounds consist of several clay pits (collection requires a permit) but have been closed since the 2013 landslide that engulfed several schoolchildren.
Amenities:	Cherokee Regional Park—the part on the top of the river bluffs—has picnic facilities.
Information:	St. Paul Parks & Recreation Department, City Hall Annex, 25 West 4th St., Suite 400, St. Paul, MN 55102. Phone: 651-266-6400. Website: www.stpaul.gov/departments/parks-recreation.

WABASHA STREET CAVES
RAMSEY COUNTY

Wabasha Street Caves offers the only cave tour in St. Paul at present. Originally a mushroom cave operated by Albert Mouchenott, a French immigrant, it was later acquired by William Lehmann, who achieved local fame as the "Mushroom King." On October 26, 1933, he opened a nightclub in the cave, dubbed Castle Royal, with a fancy patterned brick façade that you can still admire today. Mushrooms, not surprisingly, loomed large on the "dollar menu" that he touted. The chandeliers, fountains, and tapestries that graced the establishment came from the recently demolished Gates mansion, as in "Bet a Million" Gates, the largest manufacturer of barbed wire in the

Castle Royal Cave as advertised in the 1930s when it was a nightclub.
(Photo: Josie Lehmann Collection, Wabasha Street Caves)

nation. The ceiling was stuccoed to prevent the incessant rain of sand grains that would otherwise give you that gritty feel while eating. Doorways were carved out in the shape of mushrooms. In the big band era, performers like Cab Calloway, the Dorsey Brothers, Harry James and the Coronado Orchestra played there.

Castle Royal was refurbished in 1977. The venture added an Art Deco motif. The cave had a Bogart-style Casablanca atmosphere, with its 60-foot bar and dance floor. An aquarium with portholes was suspended from the ceiling, giving new meaning to drink like a fish. Khaki-clad waitresses served up the gourmet fare. Within a year, however, the owners found that the fancy menu and valet parking weren't selling well in what was dismissively called the "hamburger area" of town and they reluctantly reoriented the menu toward fast food.

Castle Royal was purchased by Bremer Construction in 1992, which originally planned to use it simply for office space and heavy equipment storage. Under its ownership, the cave flourishes to the present day as the Wabasha Street Caves. The Bremers rent out 12,000 square feet of finished space for wedding receptions and other events. The cave consists of six parallel passages—some finished, some unfinished, and some filled with debris—connected by cross-cuts. The industrious Bremers soon hosted the Great American History Theater and added gangster and ghost tours of St. Paul, a cave tour, and a coffee shop. Their success is a shining example of how the caves of Mushroom

Valley, rich in history, can be utilized and made profitable again under current economic conditions.

Right around the other side of the bluff from the Wabasha Street Caves are the giant L&M Cave entrances. They are St. Paul's version of the Griffith Park caves in Hollywood (where many cave scenes in Star Trek and other movies were filmed) in that these caves appeared on local Twin Cities television many times until they were gated, ending public access.

Directions:	215 Wabasha St., South, St. Paul, MN.
Seasons/Hours:	Year-round, but call or check website for the latest tour times. Fee.
Length:	About 500 feet of cave passages are shown on the tour.
Precautions:	N/A
Amenities:	Full-length bar for catered events, Swing Night with stage, St. Paul Gangster Tour (bus). Grumpy Steve's Coffeeshop is next door.
Information:	Wabasha Street Caves, 215 Wabasha St. South, St. Paul, MN 55107. Phone: 651-224-1191. Website: www.wabashastreetcaves.com.

YE OLD MILL, MINNESOTA STATE FAIR
RAMSEY COUNTY

This attraction is only open twelve days per year during the Minnesota State Fair, running from late August through Labor Day and it's easily the best "underground" thing to do at the fair! The ride has had remarkable longevity, having been around for more than a century, a "tunnel of love," 972 feet long, which Greg has ridden many times. More difficult to explain is the layout of the tunnels, arranged in two great loops like a tangled shoelace rather than arrayed symmetrically like a bowtie or clover leaf.

Constructed by the Philadelphia Toboggan Company in 1915, it's one of five such rides around the nation, four of which were owned and operated by the Keenan family until 2018. The first Ye Old Mill (under various name changes) has been a longtime fixture at Kennywood Park near Pittsburgh, where it began operation in 1898. It's been re-themed multiple times, including as an educational tour of the world in gondola-style boats, and as a sort of aquatic haunted house with animated scenery.

The Ye Old Mill in Minnesota has more consistently retained its original flavor. The building with the big red mill wheel (which pushes the water along) is built on the pattern of a tobacco barn, according to Jim Keenan, who provided the authors with his reminiscences. You board one of the plywood (now fiberglass) boats, while admiring the phantasmagoric painted mural which adorns the interior of the shed, usually involving

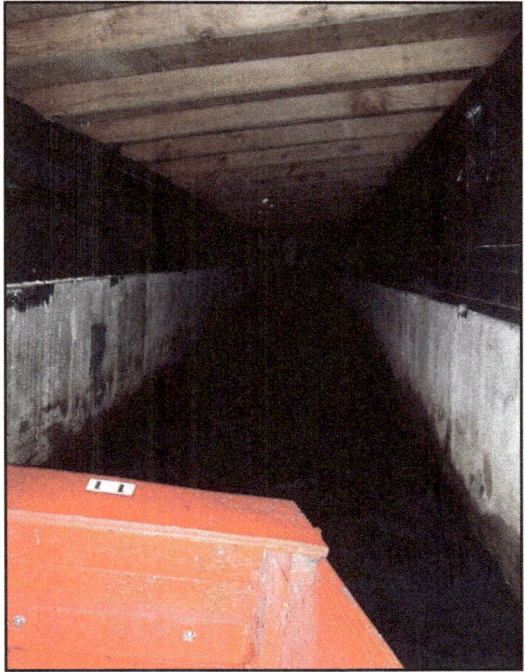

Ye Old Mill is a classic tunnel of love with a thousand feet of privacy. (Photo: Greg Brick)

Alpine scenery, polka-dot toadstools, and elves. The boat vanishes into the dark tunnel. The noise of the crowd outside quickly dies away and the sound of the flowing water and the boat bumping the wall is all you hear…unless you choose to proposition your boat mate! Most interesting, after traversing stretches of dark tunnel, you encounter several lighted dioramas like Snow White and the Seven Dwarfs, or similar fancies. Some of these scenes were repurposed from depart-ment store windows, according to the former owners. You can read reminiscences on Facebook which often involve mischievous tunnel lovers who would "soap" the tunnels, creating mounds of foam. Fully aware of this, the owners added an anti-foaming agent to the water but it still occasionally happens!

While you're at the fair, stop by the DNR Building and enjoy the mining exhibits, as well as the "What's Under Your Feet" computer display, which correlates your street address with geologic maps. These exhibits are not set out every year, however.

Directions:	On the Minnesota State Fairgrounds in St. Paul, at the corner of Carnes and Underwood.
Seasons/Hours:	During the Minnesota State Fair, 12 days in late August through Labor Day. Fee.
Length:	972 feet of boat tunnel.
Precautions:	N/A
Amenities:	The MN State Fair.
Information:	Minnesota State Fair, 1265 Snelling Ave. North, St. Paul, MN 55108. Phone: 651-288-4400. Website: www.mnstatefair.org.

POLAR BEAR DEN, COMO ZOO AND CONSERVATORY
RAMSEY COUNTY

Polar bears are known to carve out igloo-like spaces with their claws inside snowbanks, whether as maternity dens or seeking shelter during bad weather. Some hide out under snow roofs that bridge over frozen stream beds. Some of these dens have several rooms. Suitably constructed, and trapping the bear's body heat, these dens have been reported by researchers to be toasty warm. Enough to melt and form an even more insulating ice layer several inches thick, coating the walls!

Como Zoo in St. Paul has always been a marvel in that there's so much to see and it's all free! While the zoo has had polar bears for years, their newly refurbished habitat is called Polar Bear Odyssey. Polar bears frolic in the greenish pools to the fascination of visitors. Of interest here is the Polar Bear Den, where children can crawl through a bear-sized den, and even have a toy polar bear fabricated in the Mold-A-Rama machines, forging a small plastic bear on the spot for a few bucks. The Como exhibit emphasizes the serious impact that climate change is having on polar bears.

Polar Bear Den at the Como Zoo emphasizes the perils of climate change. (Photo: Greg Brick)

Directions:	1225 Estabrook Drive, St. Paul, MN.
Seasons/Hours:	April – September, 10 a.m. – 6 p.m.; October – March, 10 a.m. – 4 p.m. Free.
Length:	N/A
Precautions:	N/A
Amenities:	In addition to the zoo itself, visit the Marjorie McNeeley Conservatory with its Sunken Garden, the Ordway Japanese Garden, and Como Town Amusement Park (summer only).
Information:	Como Zoo & Conservatory, 1225 Estabrook Drive, St. Paul, MN 55103. Phone: 651-487-8200. Website: comozooconservatory.org.

SOIL CUBE, SPRINGBROOK NATURE CENTER
RAMSEY COUNTY

At Springbrook Nature Center you'll find the educational Crawl-Through Soil Cube for kids, with magnified models of common soil inhabitants such as earthworms and grubs. An instructional panel on "soil food webs" recognizes that earthworms, praised in the old agricultural literature for lightening the soil to introduce air and water, are in fact invasives, not native to Minnesota.

Nearby, a terrarium holds a large bull snake, often mistaken for a rattlesnake. Greg has encountered this scary look-alike many times while hunting for caves in Minnesota!

While the park hosts many activities and festivities during the year, a personal favorite is the Halloween festival, when lighted pumpkins are placed along the three-mile trail system

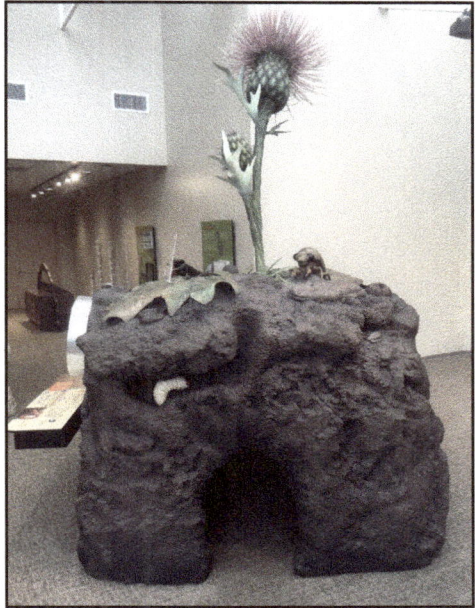

The Soil Cube displays its magnified inhabitants for children to enjoy. (Photo: Greg Brick)

in the park. Having hiked the length of the stream (Spring Brook), flowing through the park and finding some seeps but no actual springs, the origin of the name is somewhat mysterious. Nonetheless, the park survived a horrific 1986 tornado that just seemed to hover in place, wiping out many old trees.

Directions:	100 85th Ave. NE, Fridley, MN.
Seasons/Hours:	Open daily from 9 a.m. to 5 p.m. Free.
Length:	N/A
Precautions:	N/A
Amenities:	Picnic tables, three miles of gravel hiking trails.
Information:	Springbrook Nature Center, 100 85th Ave. NE, Fridley, MN 55432. Phone: 763-572-3588. Website: www.springbrooknaturecenter.org.

Overlook reveals an active granite quarry in operation. (Photo: Greg Brick)

GRANITE QUARRY OVERLOOK
STEARNS COUNTY

The best vista over an active granite quarry in Minnesota can be found here. Currently owned by the Martin Marietta Corporation, there's a small museum in a wooden shed with displays of samples and posters about quarrying. Here better than anywhere you get a feel for what replacing derrick-style cranes with drive-in truck quarries meant, as the technology of quarrying has changed. The older derrick type quarry with guywires resulted in deep pits, whereas the ramps of the newer style quarries make for elongated pits.

Called the North Quarry, it's 500 feet deep and covers 25 acres at present. The overburden of worthless saprolite ("rotten rock") must be removed when opening a new section of quarry. A lake the color of Mountain Dew fills the bottom. The granite itself likely extends for miles in depth supposing it was economically viable to go down that far. This quarry is especially renowned for its mix of colors: reds and grays, with diabase dikes providing a dark blue addition. Quarrying usually continues sideways until an obstructing diabase dike is encountered. Some of these dikes formed when the continent of North America (such as it is today) nearly rifted apart 1.1 billion years ago. Igneous rock filled the cracks.

This quarry is devoted to making aggregate, manufactured sand, and supplies railroad ballast over the Upper Midwest region. It supplied the material for a major expansion of the Minneapolis-St. Paul Airport runways. Gyratory cone crushers chew the rock down to the desired size.

Other nearby granite quarries focus on dimension stone and are run quite differently. With dimension stone, imperfections such as joints running through the rock, are detrimental. Almost no blasting takes place. Instead, large blocks are cut from the quarry face by wire saw and sent to Cold Spring Granite Company for further sawing into countertops, mausoleums, statuary, etc.

Directions:	1450 Division St. West, Waite Park, MN.
Seasons/Hours:	Check with management.
Length:	N/A
Precautions:	Access to the wheel-chair accessible observation platform is by special arrangement only, call ahead.
Amenities:	Picnic tables, viewing platform.
Information:	Dan Bokinskie, P.O. Box 7517, St. Cloud, MN 56302, Phone: 320-345-7182; www.martinmarietta.com

STEARNS HISTORY MUSEUM

"On Solid Ground," the quarry exhibit at Stearns History Museum in St. Cloud, is for people who like to know how things work. The recently renovated exhibit depicts the history of quarrying from the 19th century to 1960. Before 1900, quarrymen sometimes used dynamite to blast chunks from the bedrock, though often these pieces were too irregular to be usable. After the turn of the century, improved drills and bits helped the workers to more easily extract rectangular blocks by drilling channels into the stone. Later, they used electric charges and other power tools to facilitate the work. Derricks enabled the quarrymen to lift the large blocks, called loaves, from the quarry and place them onto trains. The museum also offers a 22-minute video, *Granite Country USA*.

Information: Stearns History Museum, 235 S. 33rd Ave., St. Cloud, MN 56301. Phone: 320-253-8424. Website: www.stearns-museum.org.

QUARRY PARK AND NATURE PRESERVE
STEARNS COUNTY

This group of 20 granite quarries making up Stearns County park (they are each assigned numbers on the park trail map) was active during the first half of the 20th century and most of them have separate historic names. Commercially known as the St. Cloud Red Granite, this dimension stone went into the construction of historic buildings, including the Landmark Center in St. Paul. The 220-acre park was opened to the public in 1998.

The quarries are numbered sequentially in roughly clockwise order around the park, each of them associated with a pile of rejected stones. Most of the quarry pits are flooded and used for swimming, scuba diving, or fishing. Despite the excellent trail system, it's a lot of ground to cover, so if you're limited for time the best place to see is the northwest corner of the park, near the main entrance gate, where the features of interest are most concentrated. There you'll find the "Historic Derrick Area," containing the sole remaining derrick with its spider and guywire construction. Really deep quarries are associated with derricks because the cranes were able to hoist blocks from deep inside the pits. Whereas most modern truck quarries are laid out on a more horizontal basis (see GRANITE QUARRY OVERLOOK).

Geologically this granite was emplaced miles below what was then the surface of Minnesota, forming the core of the Penokean Mountains, worn flat in the nearly billion years since. This has exposed their granite roots at the surface.

The quarry of most interest to geologists is Quarry 13. On the eastern shores of the quarry pool, just off the main trail through the park, you'll find a pink granite outcrop that

Quarry 13, now flooded, with granite (foreground) polished glassy smooth by the glaciers. (Photo: Greg Brick)

has been polished glassy smooth by the passage of glaciers during the last Ice Age. Glaciers carried boulders of this granite over large swaths of Minnesota, leaving them as glacial erratics.

Now look at the southern shore of the quarry pool, to your left. This is a black diabase dike, covered with pancake-sized green lichens. Diabase is a much harder and less useful rock type, so upon reaching it the quarrying stopped in that direction. Notice how the water level in the neighboring Quarry 12 is much lower. This is because the dikes are so hard and impervious that they form a good seal that the water cannot easily leak through.

If you have more time, it's worth visiting the southeastern corner of Quarry Park. At Quarry 8 (the Trebtoske Quarry) you'll find the most mountainous of the quarry piles, with a viewing platform on top, reached by stairway. The pit is correspondingly the deepest in the park. By the way, don't confuse deepest pit with deepest water. While Quarry 2 (called "Melrose Deep 7") has the deepest water (116 feet) the surface of the pool is near ground level, so it's not very dramatic, unlike the vertigo-inducing Trebtoske pit. (Hold on to the railings!)

The somewhat larger Quarry Park Scientific and Natural Area (SNA) covering 323 acres adjacent to the south of the park proper, allows you to see what the area looked like before quarrying began. Instead of the pit-and-pile landscape of the park, with its white birch trees, you'll see low granite domes in an oak grove. On some of the domes you can find cactus owing to the desert-like dryness of the rock outcrops. Other unique aspects of this ecosystem can be seen on the self-guided Eco-Walk. It's a long loop, however, and you won't see much more than you already have, so skip it if you're short on time.

Directions:	1802 County Rd. 137, Waite Park, MN 56387.
Seasons/Hours:	Year-round. 8 a.m. to 30 minutes after sunset. Vehicle parking permit required.
Length:	Trails run for thousands of feet.
Precautions:	Rugged trails, cliffs, and unstable quarry piles.
Amenities:	Picnic shelters, swimming, fishing, mountain bike trails, rock climbing, lighted cross-country ski trails, scuba diving.
Information:	1802 County Rd. 137, Waite Park, MN 56387, 320-255-6172, www.co.stearns.mn.us/Recreation/CountyParks/QuarryParkandNaturePreserve.

View of the Minnesota bank of the St. Croix River, as seen from Wisconsin's Interstate State Park. (Photo: Doris Green)

ST. CROIX RIVER VALLEY

T he St. Croix River valley was carved as the last Ice Age ended, and so its natural caves lack the superlative antiquity of those in southeastern Minnesota—tens of thousands of years, compared with hundreds of thousands. By contrast, the valley's early lumbering days, which predates settlement elsewhere in the state, give its underground aspect a unique cast.

JOSEPH WOLF CAVES
WASHINGTON COUNTY

According to local legend, French trader Jules St. Pierre opened a trading post at the location of a spring in the Jordan Sandstone bluff in what is now Stillwater, Minnesota, in 1838. By 1868, a Swiss brewer named Joseph Wolf was hollowing out a space around the spring, enlarging it to form a natural refrigerator for the beer he was brewing. But the brewery shut down during Prohibition and the cave sat vacant for decades.

Joseph Wolf Caves played a role in the transition of Stillwater from lumbering to the tourist destination it is today. (Postcard from the Gordon Smith Collection)

In 1945, local entrepreneur Tom Curtis purchased the vacant cave, enlarging it, adding a trout pond fed by the natural spring water and a boat tour through a canal system, which opened in 1958, running until 1973. This attraction, at first called Curtis Caves, then Stillwater Caves, signaled the transition of Stillwater from an industrial town, based on sawmilling, to the tourist destination it is today. During the Cold War, the cave doubled as a fallout shelter and was stocked with Civil Defense supplies. This became the only Minnesota cave where actual Cold War ceremonies were enacted, with schoolkids being marched into caves. That Cold War mascot, "Bert the Turtle," famous for his admonition to "Duck and Cover," had now acquired a rocky carapace.

Under the name *Blue Grotto*—recalling the famous Blue Grotto on the Isle of Capri in Italy—the cave served as a mood piece for an adjacent Italian restaurant. At other times, keeping with the Italian theme, the Venetian type canals inside the cave were emphasized. The cave itself was intermittently reopened for tours after Tom Curtis sold the operation, most recently as the Joseph Wolf Brewery Cave, when it was elaborately decorated for Halloween. After several years of remaining in limbo, the caves will soon be reopening as part of the Lora Hotel-Feller Restaurant complex.

Directions:	Lora Hotel, 402 Main St. South, Stillwater, MN.
Seasons/Hours:	N/A
Length:	Hundreds of feet.
Precautions:	N/A
Amenities:	Lora Hotel-Feller Restaurant complex. Many other shops and restaurants nearby.
Information:	Lora Hotel, 402 Main St. South, Stillwater, MN 55082. Phone: 651-571-3500. Website: www.lorahotel.com.

BOOM SITE CAVE
WASHINGTON COUNTY

From 1856 to 1914, the "Boom Site" on the St. Croix River, north of Stillwater, was the location of a lengthy boom stretched across the river, catching logs coming downstream from the pineries, where they had been cut the previous winter. Here, the logs were sorted to determine ownership according to the distinctive company marks struck into the wood. All that work required laborers and camps to house them. That's where caves enter our story.

According to signage at the Boom Site wayside, there was "a large cook house" for laborers in the logging industry. "A cave located 30 feet below the cook house was used as a storage cellar for food supplies. An elevator shaft connected the two."

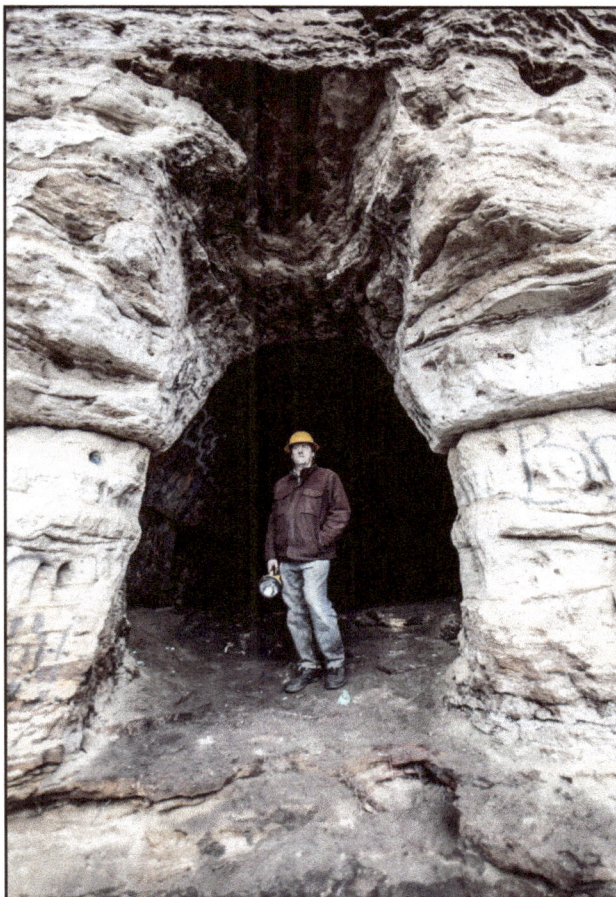

Greg Brick at the entrance to the Boom Site Cave, a lumberjack's food storage cellar. (Photo: Tony Andrea)

Dug in the Franconia Sandstone, the artificial cave measures 66 feet long, 16 feet wide, and six feet high, and is easily accessible from the highway pullout. Walk down the steel stairway to the St. Croix River, turn right at river level, and look for the large open entrance about 50 feet to the south along the sandstone outcrop. Beyond, a series of seven springs emerge from the sandstone.

The cave's walls are thickly covered with graffiti and at the back end you'll come to a concrete block wall of more recent construction. Kids have already dug under the wall, as the cave was rumored to stretch a great distance, but having gone back there recently, crawling under the wall, Greg found what must have been the collapsed elevator shaft. There was barely enough room to turn around amidst the debris.

Having explored the cave, there are other places of geological interest at the Boom Site. The steel stairway you descended adjoins a short sandstone canyon; its rock walls display colorful cross-bedding. And if you hike north along the sandstone outcrops about half a mile over a rollercoastering lover's lane type trail you'll eventually come to a secret waterfall.

Directions:	Drive 2.5 miles north of downtown Stillwater on State Highway 95. NOTE! There are four vehicle pullouts, all of them part of the extended "Boom Site." The southernmost is the Public Water Access (for boats), the next is the Scenic Overlook (with geological marker), but it's only the third (Historic Marker) and fourth (Wayside) that will place you in close proximity to the cave itself.
Seasons/Hours:	Year-round, dawn to dusk.
Length:	66 feet.
Precautions:	Lengthy steel stairway to get to river level.
Amenities:	Picnic tables.
Information:	St. Croix National Scenic Riverway, 401 Hamilton St. North, St. Croix Falls, WI 54024. Phone: 715-483-2274. Website: www.nps.gov/sacn.

GEOTOURISM IN MINNESOTA

Geotourism is a new word for an old activity, tourism with a geological emphasis. The core of geotourism is curiosity about geological phenomena, whether at the casual, amateur, or professional level, beyond the level of merely aesthetic appreciation. In the ancient world, the Greek writer Herodotus (fifth century BC) arguably became the world's first geotourist and perhaps the Father of Geotourism when, on a trip through Egypt, he described the accumulation of sediments in the Nile River delta. Among the Romans, Pliny the Elder died while observing an eruption of Vesuvius—the first geotourist martyr.

The Geological Society of Minnesota (GSM) is an organization of amateurs and professionals interested in geology. Since the 1950s they have installed bronze markers around the state at points of significant geological interest. In the past several years, all of the more than 70 statewide geology markers have been revisited, some of them revamped, and an online interactive map prepared, posted at: http://roadmarker.geosocmn.org/

KNAPP'S CAVE
WASHINGTON COUNTY

A gaping cave entrance in the reddish sandstone cliffs has attracted generations of canoeists on the St. Croix River to stop and investigate. Geologist R.W. Strong began touting this "mammoth cave" of the St. Croix valley back in 1900, and despite the hyperbole of the comparison, it's recognizable today. Knapp's Cave remains the largest natural cave in the St. Croix River valley and the only one with significant archeological remains. The namesake Captain Knapp was a steamboat captain out of Osceola, Wisconsin, just across the river. While some claim that the cave was an overwintering shelter for new immigrants to the valley, a lack of the debris that such habitations leave behind makes the claim unlikely.

The archeologist Lloyd Wilford excavated Knapp's Cave in 1951 and found it to be a temporary campsite of the Woodland period. He called it "Leslie Cave" after the contemporary landowner, but it's marked as Knapp's Cave on U.S. Geological Survey quadrangles. It was later explored by members of a local caving club, the Swedish Underground, this area having been settled by Swedish immigrants. Being floored with thick, dry reddish sand, it was an ideal party cave, and the Cedar Cliff residents have had to restrict access to the waterfront from the landward side, but the shoreline remains public.

Knapp's Cave fronts the Cedar Bend country of the St. Croix River Valley. (Photo: Tony Andrea)

As you enter Knapp's Cave, with its beach-like floor of soft, reddish sand, look up into the spacious vaults ahead. You will see upper level passages that you can enter by scrambling up the boulders. This upper level is called "Spider Haven" and certainly lives up to its name. The spider is *Theridion tepidariorum*, which is the common house spider, but it grows to a disturbingly large size. If you make it past those gatekeepers, the passage tapers down to tight crawlways that past cavers have obviously attempted to dig open. The speculation is these passages eventually lead to sinkholes in the fields beyond, a route for the water that flushed out the cave.

Directions:	Accessible by boat on the St. Croix River, 3.5 miles south of the CR 243 bridge at Osceola, WI. Paddling downstream (south), begin watching the Minnesota shoreline when you come around Cedar Bend and a railroad bridge comes into view. Come ashore at the gravel bar projecting slightly into the river, where you'll also see a spring of water. Follow the trail uphill through the woods. UTM coordinates for the cave are 518537 E, 5014718 N.
Seasons/Hours:	Year-round, but fall leaf season is the most scenic. Free.
Length:	300 feet.
Precautions:	Do not approach the cave by parking in the Cedar Cliff community on the landward side, as the residents have strong objection to this and may try to have you towed.
Amenities:	As part of a leisurely canoe trip.
Information:	St. Croix National Scenic Riverway, 401 Hamilton St. North, St. Croix Falls, WI 54024. Phone: 715-483-2274. Website: www.nps.gov/sacn.

Bake Oven Cave, Interstate State Park
Chisago County

The Mid-Continent Rift nearly tore what is now North America in half about 1.1 billion years ago, until being stopped by an even greater force—the landmass of Europe crunching up against us, in a plate tectonic grind. As the rift opened—stretching from Michigan to Kansas—vast quantities of flood basalt lavas emerged, flow upon flow. You can see these layers—seven or ten of them, according to the authority consulted—stacked up in the staircase landscape of Taylor's Falls, MN. And you can examine them up close at Interstate State Park.

The best way to get a grasp of what's going on geologically—after digesting the bronze geological marker in the park—is to walk downhill on the asphalt road to the lower steamboat landing and proceed back uphill, rather than going pell-mell among the outcrops as most visitors do. Because down there you'll be able to perceive the individual lava flows, about 20 feet thick, giving you a better notion of the basic structure. The tops

A view of sky from the depths of Bake Oven Cave. (Photo: Greg Brick)

of the flows are marked by vesicular (pock-marked) basalt—the dark rock making up the flows. In these lava flows you'll see pipe vesicles where gas trapped in the lava tried to bubble to the surface.

Some features along the self-guided Glacial Potholes Trail have sign boards naming them. Make your way uphill to the Bake Oven, a pothole entered by means of a steel stairway, providing what is nearest to a cave experience at the park. Shaped like an oven but actually cool inside, you can look up and see open sky, with vines dangling down. In the walls of the pothole you'll see the spiral fluting which, in extreme cases, has led to potholes being compared to gigantic snails. Notice several iron seeps staining the walls a rust red color. There's a lot of iron in basalt, which is also why it's such a heavy rock.

But the Bake Oven is not even the largest of the potholes. Continue uphill among the flows to the Bottomless Pit, more than 60 feet deep and 12 feet wide. Nearby is the Lily Pond, a pothole that was maintained as an actual lily pond years ago. Considering that water lilies root in six feet of water, this pothole is not anywhere near as deep as its neighbor. Sometimes the potholes really snuggle up: continue through the Devil's Parlor, the rocky rift arched over by a stone bridge, where you'll see potholes so close together that they've coalesced.

How did the 80 potholes form? In pioneer days, it was assumed that they were Indian cooking pots—a theory that would certainly justify the Bake Oven's name! Later, it was assumed that potholes were formed by meltwater plunging down through crevasses in the glacier or by natural mill stones, or "grinders," swirling around in the mad rush of postglacial melt rivers, grinding the bedrock away. But H. S. Alexander, a scientist at Macalester College in St. Paul, published a laboratory study in 1932 proving that the potholes were formed by sand-size abrasion, not the boulder-sized grinders. He constructed an artificial concrete pothole armed with a jet of water that could be directed at various angles to observe abrasion in action as he fed the jet with various particle sizes. He concluded that the abrasion was due to eddying, not plunging water, and that it was the sandpapering effect that did most of the work. The Visitor's Center has a nice display on grinders or grindstones and corrects previous misconceptions of how the potholes formed.

After the pothole forming phase, the boisterous glacial river cut even deeper into the old lava flows, creating the St. Croix Dalles, leaving the potholes high and dry. By the time you reach Angle Rock—a high perch overlooking a right angle bend in the river, from south to west—you gain insight as to why the potholes formed on the tight inside bend of the river, where eddies abound, rather than on the Wisconsin side, where the potholes are smaller.

When you complete the Glacial Potholes Trail, there's a separate southern unit to the park that will have some interest. Behind the park office, you'll find a hiking trail leading through a culvert under U.S. Highway 8 to Curtain Falls. The wooden stairway on the other side is long and steep. Here the colorful sandstones are only half as old as the basalt on which they rest. They show good examples of cross-bedding, like the cross-section of a sand dune, indicating they were formed in the same way.

The park itself was established in 1895 in a joint effort with Wisconsin and was the nation's first interstate park. This park has a twin across the St. Croix River, Wisconsin's Interstate State Park, which is the western terminus of the Ice Age Trail. The visitor's center has a stuffed mammoth. A park sticker from either state permits access to the park on both sides of the river.

Directions:	From the Twin Cities, follow I-35 north to U.S. 8. Take U.S. 8 northeast to the park.
Seasons/Hours:	Year-round, 8 a.m. to 10 p.m.; however, naturalist programs and pothole tours are available only during the summer. Park fees.
Length:	The stone surfaced Glacial Potholes Trail is a relatively easy quarter mile walk, although it includes several stone stairways.
Precautions:	Cliff edges with steep drop-offs.
Amenities:	State park visitor's center and the nearby scenic town of Taylor's Falls. Picnic shelters, volleyball, camping, hiking, canoeing, fishing, rock climbing.
Information:	Interstate State Park, P.O. Box 254, 307 Milltown Rd., Taylors Falls, MN 55084. Phone: 651-465-5711. Website: www.dnr.state.mn.us/state_parks/index.html and use the Park Finder.

CRYSTAL CAVE
PIERCE COUNTY, WISCONSIN

From the beginning, Crystal Cave has offered adventure to people seeking to explore the earth and test themselves against rock and mud, to rediscover their own inner and physical strengths. Located east of the Twin Cities in Spring Valley, Wisconsin, it still offers a rugged, exploratory tour for adventurers as well as a more relaxed circuit on paved walkways and ramps. Both trips feature geology and history lessons that you can experience firsthand.

Story has it that Crystal Cave was discovered in 1881 by a teenage boy chasing a woodchuck about a half mile from his farm home. When the woodchuck disappeared down a sinkhole, the boy probed the hole with a stick. When he dropped the stick and it disappeared in the depths, the boy, William R. Vanasse, went home to tell his brother, George.

About an hour's drive from St. Paul, Crystal Cave is Wisconsin's longest. (Photo: Crystal Cave)

The next day, the two boys brought a rope and a lantern to the spot and lowered themselves down a "slide" into a clay-and debris-filled room. From there they dropped into the main room on the second of the cave's three levels. In several directions, the boys saw shallow entrances to clay-filled galleries on the upper level.

For decades, Sander's Corner Cave, as it was then called, remained partially filled. Adventurers occasionally visited and explored, but no one realized the size and potential of the Vanasse find—until Henry A. Friede came on the scene in 1941.

An advertising agency manager and amateur geologist from Eau Claire, Friede had been studying caves in the area in hopes of finding one equal to the Cave of the Mounds, discovered two years earlier. He decided that of the several possible sites in the region, the most potential lay in the development of Sander's Corner Cave. In November 1941, a crew began removing the silt and debris, largely completing the job by April of the following year.

Work began on an entrance building, but even before it had been completed, Friede renamed and opened Crystal Cave to the public in June 1942. About four thousand visitors showed up on opening weekend. The new name referred to the quartz crystals found throughout the cave. Since then, more than one million visitors have toured Crystal Cave, which has been further cleaned and explored.

The cave has changed ownership several times. Previous owners Blaze Cunningham and Jean Place opened the cave and surrounding property to exploration by the Wisconsin Speleological Society and Minnesota Speleological Survey, leading to the discovery of additional passages in Crystal Cave and adjacent caves. Today Crystal Cave is Wisconsin's longest, with about 4,600 feet of surveyed passages.

Current owners Eric and Kristin McMaster purchased the cave in 2012 and have continued an emphasis on education combined with entertainment. When visitors walk from the parking lot to the limestone cave building, they pass eight framed displays about cave geologic periods. Another outdoor display describes the property's native pollinator garden, and in-cave displays educate about such topics as fossils and the bats of Crystal Cave.

Located on the edge of the Driftless Area, Crystal Cave lies within dolomite (magnesium-rich limestone) and features iron ore, chert, and, of course, quartz crystals. There are also two snail-shaped fossils about the size of a nickel—both gastropods from when the area was part of a giant inland sea.

Visitors also see the Ballroom, the cave's largest chamber, where bats hang out in winter and where more than one couple has exchanged their marriage vows. The Ghost Room features an unusual "face" in the walls and is often the kids' favorite. Most kids also enjoy Charlie's Room, where guides weave a spell by telling such tall tales as that of Cave Man Charlie, who once lived in the cave and participated in boat races on an underground river. In the dim depths, with the fantastical formations of Charlie and the remainders of his boat clearly visible in the rock, even adults almost believe in the myth. Underground, it certainly seems more plausible, than, say, a blue ox.

The walls of the Wish Room are composed of illite, a mica-like clay mineral. Clays containing illite are used for makeup, since they absorb facial oils, as well as in construction, china, and medicines. Visitors were once invited to place a coin in the walls and make a wish. The longer their coin stuck to the wall, the more likely it was for their wish to come true. Considering the numerous coins lining these walls, many visitors have had their dreams fulfilled!

Currently the guides use the coins of the Wish Room "to talk about how conservation has changed over the years," Eric McMaster said. Today visitors to any newly opened cave passages would never be invited to stick coins into the walls—any more than they would be welcome to sign their name to flowstone or break off a stalactite as a memento of their visit.

McMaster has partnered with Santa Clara University and the Wisconsin Department of Natural Resources on bat research, including surveying and banding bats. One study involved attaching tiny radio transmitters with an antenna to track individual animals. Another used ultraviolet dust to determine which bats touched other bats, leading to potential spreading of white nose syndrome.

Ecology efforts extend to the surface, where a seven-acre field near the property entrance is being developed as a pollinator prairie in partnership with the DNR and U.S. Fish and Wildlife Service. Each spring over 500 maple trees are tapped; waist-high plastic piping carries the sap downhill to collection tanks. The resulting Crystal-Cave branded maple syrup is for sale in gift shop.

On the surface there is 18-hole Tee-Rex minigolf. Guests learn about different dinosaurs as they meander through the course. There is even a giant T-Rex fossil skull to putt through and dinosaur eggs to putt around.

Adventurous visitors can make reservations to take a two-hour wild cave trip through **South Portal Cave**, hiking down the ravine and entering the cave through a rough wooden door.

South Portal Cave has tight passages, no lights, and an unimproved path. A registration fee includes boots, a hard hat, and headlight. This tour begins with a caving orientation session. Minimum age is 14, and children 14 to 18 must be accompanied by an adult. Reservations are necessary and there is a maximum of six people per tour. Participants need to sign a liability waiver, be okay with tight spaces, and be physically fit. See Doris Green's 2019 book *Wisconsin Underground* to read about 75 other subterranean sites in Minnesota's neighboring state.

Directions:	Located about one mile west of Spring Valley on State Highway 29, approximately 50 miles east of St. Paul
Seasons/Hours:	Open daily: April through May, 10 a.m. through last tour, which departs 4 p.m.; Memorial Day weekend through Labor Day, 9:30 a.m. through last tour, which departs 5 p.m. daily; and September through October, 10 a.m. through last tour, which departs 4 p.m. Fee.
Length:	Tour lasts about one hour.
Precautions:	Light jacket recommended; temperature is 50 F year-round. For safety, strollers or backpack style baby carriers are not allowed in the cave; however, soft, front style baby carriers are permitted. Small purses and camera bags may be brought into the cave; lockers are available for larger items.
Amenities:	Picnic tables, restrooms, gem panning, mini-golf and gift shop.
Information:	Crystal Cave, W965 State Hwy. 29, Spring Valley, WI 54767. Phone: 715/778-4414. Website: http://www.acoolcave.org/.

NORTHERN MINNESOTA

Northern Minnesota is unlike the rest of the state's underground regions because it lacks soluble rock types like limestone and the colder climate is less favorable for the sort of weathering that produces lengthy limestone caves. But it ranks first in terms of mining related features to enjoy.

In the northeast in particular, there is no karst, no limestone or dolostone. When you drive northeast in Minnesota, mountains replace level or gently undulating land and forests overwhelm fields.

Several glaciers advanced, retreated, and advanced again over northeast Minnesota, scraping and scouring the land. There is little glacial fill here, since most glacial debris was deposited along the edges of the thick ice sheets, particularly in their terminal moraines. When they melted, the glaciers left exposed much Precambrian rock.

These rocks include granites formed deep within the earth's crust, basalts marking ancient lava flows, and sandstones born in prehistoric shallow seas. They also include the iron formations of Minnesota's Cuyuna, Mesabi, and Vermilion ranges, stretching in a diagonal swath across northeast Minnesota from Crosby to Ely.

While each range has played an important role in the state's mining history, their ore formations are not identical. In the Vermilion Range, for example, the ore bodies tend to extend vertically, like upright bean pods buried underground. In contrast, the ore bodies in the Mesabi Range tend to lie within 500 feet of the surface and to spread laterally over a wide area.

Iron is found in a variety of rocks within these formations. Much of the highest-grade black-red hematite, consisting of 65 percent iron, has been mined; however, technology advancements have made mining the less iron-rich taconite profitable. Taconite contains magnetite as well as hematite, and modern production methods now separate the iron magnetically. Even though iron is the fourth most common element of the earth's crust (after oxygen, silicon, and aluminum) it's usually not found in the high concentrations present in Minnesota's iron ranges.

Ore was first shipped from the Vermilion Range in 1884, though it had been discovered years earlier during the gold explorations of the 1860s. In their quest for gold, prospectors initially ignored the mineral that would prove more valuable to Minnesota over the long term. Less than a decade after the first shipment of iron ore left the Vermilion Range for Two Harbors, the first iron ore shipment left the Mesabi Range; the first iron ore was shipped from the Cuyuna Range in 1911.

The majority of the sites in this section commemorate the state's iron mining and processing, although one listing marks the Rainy Lake Gold Rush along the Minnesota-Canada border and others document manmade and natural sites along the North Shore, around Duluth, and farther south at Banning State Park.

Yet the iron mines, especially the pit mines of the Mesabi Range, have had the most widespread impact on the landscape. Surface mining created many gargantuan pits. Some of these have been reclaimed as lakes and are today popular among scuba divers and fishermen. Surface mining also created huge tailings piles, which in many places have been planted with trees and now appear as young forests. Not since the glaciers crunched across these lands has the topography been so altered.

CROFT MINE HISTORICAL PARK
CROW WING COUNTY

Minnesota's sometimes forgotten Cuyuna Range lies southwest of the better known Mesabi and Vermilion ranges, hidden beneath tracts of lakes and woodlands. Yet its story is worth remembering and the Croft Mine Park is worth a stop the next time you're heading to the Brainerd Lakes area.

Situated on the original mine site, the 17-acre park contains a museum, mining equipment, miner's cottage, and a tour through a mine replica that's as real as advice from retired miners can make it. The simulation, finished in 1988, provides a first-hand experience of the sometimes dangerous business of mining underground on the Cuyuna. The only thing you don't do is pick up a hammer or drill or heave the ore into a waiting tramcar.

Iron mining began later here than in the Mesabi and Vermilion Ranges. Cuyler Adams noticed a magnetic pull in the area while determining property lines; he organized the Orelands Mining Company in 1903 with William McGonagle and William C. White. Cuyler's dog was named Una—hence the Cuyuna town name. The first mine in this range was the Kennedy Mine near the town of Cuyuna, and soon other mines opened in the area.

Discovered on a Swedish immigrant farm, the Croft Mine's owners dug its original shaft to a depth of 634 feet deep. The mine began shipping ore in 1911, and operated until 1934. During that span, the mine was run first by the Merrimac Mining Company, followed by the Youngstown Mining Company from 1928 to 1931, and finally by the Hanna Mining Company. Over the years, production totaled 1.77 million tons of ore. During its heyday in 1920, the Croft Mine employed 165 men.

Now part of the Cuyuna Country State Recreation Area, the Croft Mine Historic Site serves as a stop on a 27-mile long mountain bike trail system, covering nearly 800 acres from Yawkey Mine Lake, on the east, to Huntington Mine Lake on the west. The trails pass several former mining pits and rock stockpiles, not to mention 5,000 acres of regenerated vegetation.

Today, you can wander about the 17-acre Croft Mine Historic Site to view large mining equipment, a dynamite "house" resembling a short silo, and a honeymoon row cottage built by George H. Crosby, founder of the town. The white cottage—the last one not remodeled and still remaining of a several-block section of row houses—was moved to the park when it was developed. The only original building still standing in the park is the dry house, where the miners changed their clothes before and after their shifts underground, and which today houses the park's museum.

Weekend guided tours begin at the ticket booth and wind above ground on a path that lies directly above one of the underground tunnels. The tunnels themselves, along with the mineshafts, are now flooded. Walking along the mowed path, you can see the sites where the hoist house and shaft house once stood, narrow gauge tracks where ore trams once ran, and a side trail leading to a beaver pond. The pond was created when a sinkhole developed following dynamiting in the mine.

From L to R: man cage, shaft house, and dry house at the Croft Mine on the Cuyuna, Minnesota's forgotten iron range. (Photo: Greg Brick)

Alongside the ramp leading into the reconstructed shaft house, you can spot ore buckets and a two-and-one-half ton skip, or ore cart. In the shaft house, a modern elevator waits to take you "underground." Originally, steam engines powered the elevator and air and water pumping systems.

When the elevator comes to a stop, the doors open into a realistic replica of a drift, or horizontal tunnel. The drift owes its authentic appearance to carefully crafted cement and fiberglass "rock" walls of the same dimensions as the real thing. Still, the tunnel's comforts would have surprised the miners who once worked 10-hour shifts in the Croft. The cement floor is smooth underfoot, water does not drip from walls and puddle on the floor, and electric light provides plenty of illumination. The temperature is 60 degrees, a full 10 degrees warmer than when the real mine was in operation.

Thanks to the steady, above freezing temperature of the Croft, it was mined all year round. In the winter the miners stockpiled the ore for shipment during warmer weather.

Cuyuna Range miners had to dig through more than 110 feet of glacial drift to get to the iron ore. They would sink a shaft into the adjacent slate and other more stable rock, and then dig drifts to the softer ore bodies.

At Croft, miners constructed two shafts and six drifts. They used one shaft to bring workers and ore in and out of the mine; they carried timbers into the mine through the second shaft. As the miners dug into the soft ore, they used timbers to shore up the cavities they formed.

In the replica drift, you can see a small dog hole, just large enough to permit a miner to crawl in and collect a rock sample. You also can see other passages. Digging upward, the miners created diagonal passages into subdrifts where much of the work was done.

When digging in the soft ore body, the miners would often work a "contract" or circle, mining one pie-like section at a time. They would shore up each wedge as they excavated it. When a contract or passage had been mined out, the miners would remove the timbers and allow the passage to collapse behind them. They would then reuse the timbers in a new section.

The ore in the Croft Mine had 55 percent iron content, the highest on the Cuyuna Range. The ore was moved by pushcart in the mine, then hoisted to the surface and loaded into narrow gauge rail cars for transport to a public rail line.

Also visible in the replica drift is a first aid room furnished with aspirin, bandages, and a stretcher. Emergency evacuation was on a timber cart through the timber shaft, with the cart knocking against the walls of the shaft on the slow jerky ride to daylight.

The miners used a variety of tools over time.

In the early days, most work was done by hand. One miner would hold a long handle with a drill bit at the end against the rock wall, while another miner hammered the end of the long handle. The miner holding the drill would regularly rotate the tool and be careful to keep his hands out of the way of the other fellow's hammer in the dim candlelight.

During World War I, improvements were introduced, for example, the miners switched to carbide lamps, which provided better light than candles. As at the Soudan Mine, the Croft miners used a tugger to power a scraper, which dragged the ore along the passage to the chute where it would drop into a cart for transport to the main shaft. There's even a Longyear steam-driven core drill here, used to analyze the rock and determine the direction of the tunnels.

At the end of the tunnel replica, you pass an engineer's office, desktops spread with tools and papers. The engineer determined where and how to proceed as the maze of underground passages grew.

The exit of the tunnel replica serves as one entrance to the park's museum, which showcases an array of mining tools, clothing, ore samples, and other artifacts. One of the most mesmerizing displays is that of the 1924 Milford Mine disaster. The nearby Milford disaster killed 41 miners when a section of the mine collapsed, sending a crush of mud and water into its passages. In half an hour, the Milford was completely flooded. Only seven men working in the Milford's upper levels escaped as a swamp and pond caved into the mine.

Located at the old mine site about four miles north of Crosby, the **Milford Mine Memorial Park** commemorates the disaster. The park includes the old mine shaft and features a 450-foot boardwalk over Milford Lake, which filled in after the mine closed. Inscribed on the boards are the names of the miners who died and those who survived. The boardwalk leads to the former Milford company town, where building foundations have been excavated and posted with explanatory signage.

The Croft museum tells another Milford story, too, that of Vicenzo Rocca, a miner who quit the Milford two weeks before the collapse. Tired of working in waist-deep water at the Milford, Rocca feared a potential catastrophe and went to work at the Croft. He was right to do so. The collapse occurred just 15 minutes before quitting time at the end of the day shift on February 5, 1924.

Directions:	From the Twin Cities, follow U.S. 169 north to State Highway 18 along the western shore of Mille Lacs Lake. Take 18 west to State Highway 6. Turn right (north) on 6 and drive to Crosby, through the intersection with State Highway 210 to the entrance, located about 0.5 miles north of 210. The park is also accessible from Second Avenue, eight blocks north of 210.
Seasons/Hours:	Open from 10 a.m. to 6 p.m. weekends, Memorial Day weekend through Labor Day. Off-season by appointment. Free. Accessible
Length:	Allow at least an hour or two to explore the park, take the tour, and see the museum exhibits.
Precautions:	Wear walking shoes. The temperature in the mine replica is sixty degrees.
Amenities:	Museum, library, restrooms, picnic tables, playground equipment.
Information:	Cuyuna Country State Recreation Area, 307 3rd St., Ironton, MN 56455. Phone: 218-546-5926. Website: www.dnr.state.mn.us/state_parks/index.html and use the Park Finder.

SPACEMAN OF THE IRON RANGE

In a way, the space age began at the bottom of a deep pit mine in the Cuyuna Range. On August 20, 1957, Air Force Major David Simons, donned a spacesuit and climbed into an aluminum capsule for his balloon ascent from the bottom of Portsmouth Mine. Located at Crosby in the Cuyuna Country Recreation Area, the 450-foot-deep mine protected the fragile balloon from surface winds as it filled with helium and rose to the stratosphere.

Simons reached 102,000 feet to set a record for balloonists. During his 32-hour flight, he underwent numerous medical tests as part of a pre-NASA program called Project Manhigh. Simons landed in an eastern South Dakota field when a thunderstorm interfered with plans to reach Montana.

Two other Manhigh balloon flights were completed, in June 1957 and October 1958, though neither topped 100,000 feet. But Project Manhigh lost altitude late in October 1957 after the Soviet Union launched its Sputnik I satellite and sparked the Cold War space race. The United States soon turned its focus to the development of NASA and the Mercury manned space flights.

You can view Simons' aluminum capsule (aka balloon gondola) at the National Museum of the U.S. Air Force at Wright-Patterson AFB, Ohio, and see a photo of Simons and drawing of a man crammed inside at www.nationalmuseum.af.mil (search for Project Man High Gondola). You can also visit the Portsmouth Mine, though now it's the Portsmouth Mine Pit Lake—complete with stocked fish, swimming area, and campground.

SPEAKEASY TUNNEL, BOVEY
ITASCA COUNTY

Ghosts walk in Bovey, Minnesota, throughout its historic downtown, beneath its streets, and especially within the walls of Annabella's Antiques and Café. Originally a mercantile and then a longtime hardware store, the building that houses Annabella's likely once hosted more clandestine activities. A steel door in its lower level could slide shut to hide the likely speakeasy beyond, and a series of tunnels led to the street, the city hall, and a former blacksmith shop.

The door to the tunnel entrance today leads from the recreated speakeasy in the antique mall's lower level where the mall's collection of historic treasures takes visitors back to the Prohibition Era. So does the 2017 Emmy-nominated episode of the *Gems of*

Tunnel entrance from speakeasy in lower level of Bovey antique mall. (Photo: Doris Green)

Itasca YouTube series. You can rent the speakeasy for your next theme party, or contact Annabella's for information on speakeasy tours, which include a walk through the tunnel.

Down the street from Annabella's is the former photography studio of Eric Enstrom, famous for his 1920 photograph of itinerant peddler Charles Wilden. When Wilden visited his photography studio, Enstrom posed the white-bearded Wilden sitting with his elbows resting on a tabletop, hands folded, and head bowed. Before him on the table were a loaf of bread, bowl of gruel, knife, thick book, and glasses. The resulting classic photo, *Grace*, was named Minnesota's state photograph in 2002. A huge replica of the photo today adorns the building of Enstrom's former studio.

Directions:	From Grand Rapids, follow U.S. 169 for seven miles to Bovey. In the historic downtown, turn left on Second Avenue and then right onto Second Street.
Seasons/Hours:	Open daily, 10 a.m. to 5 p.m. Call for speakeasy tunnel tour reservations.
Length:	N/A
Precautions:	Stairs to lower level and tunnel.
Amenities:	The Itasca County Historical Society center at 201 North Pokegama Ave. in nearby Grand Rapids offers exhibits on the area's history, including logging, mining, and more on the life of Eric Enstrom.
Information:	Annabella's Antique Mall and Café, 407 2nd St., Bovey, MN 55709. Phone: 218-245-2055. Website: http://www.annabellasantiques.com/.

COLERAINE HISTORIC WALKING TOUR

Pick up a free brochure describing this walking tour at the Coleraine City Hall, next to the high school, or at the Depot Visitor Center in Grand Rapids. A booklet providing more detailed descriptions of each property on the tour is also for sale at the City Hall and the Coleraine Public Library.

Named for Thomas Cole, president of U.S. Steel's Oliver Mining Company, Coleraine differed from the typical mining town in several respects—thanks largely to the efforts of John C. Greenway.

U.S. Steel hired Greenway, a Rough Rider and friend of President Theodore Roosevelt, to expand mining operations in the area and develop a company town. But Greenway's idea of a company town was far more progressive than most. He replaced tract housing, where rows of identical homes lined up like soldiers, with homes as individual as the people who eventually inhabited them.

Coleraine's walking tour passes a company housing district, as well as the general mining superintendent's home, a log lodge, and the Coleraine ski jump site. Greenway and U.S. Steel built the wooden jump for the town in 1909-1910.

To view a more typical mining town, head northeast on U.S. Highway 169 to the adjacent town of Bovey. Drive through its streets to find several rows of identical miners' homes and boarding houses, where one bed was sometimes rented to three different men, who slept in shifts—an arrangement known as "hot sheeting."

Directions: From the Twin Cities, follow U.S. 169 north to reach Coleraine, just north of Grand Rapids.

Website: www.cityofcoleraine.com/historical-walking-tour .

HILL ANNEX MINE STATE PARK
ITASCA COUNTY

The Hill Annex Mine State Park trolley provides wide views of a string of mine pit lakes. (Photo: Doris Green)

Currently a state park, the former Hill Annex Mine on the western Mesabi Range faces an uncertain future. Legislators and local leaders continue to negotiate ways to fund the park and keep it open despite low visitor revenue.

Mining ceased here in 1978, and the site was placed on the National Register of Historic Places in 1986. Two years later it became a state park.

But Hill Annex Mine differs from Minnesota's other state parks. There are no recreational trails, no map to help you hike through the park. The park office is located in the miners' two-story clubhouse, built in the 1930s.

The clubhouse contains mining exhibits, as well as the park office, which shares the main floor with displays explaining mining operations, miners' equipment, and how mining changed over time. Upstairs, you will find a historic mineral analysis lab, a geologist's office, the mine superintendent's office, a typical miner's bedroom, and even a schoolroom where the miners' children learned to read, write, and cipher.

Aside from the Mesabi Bike Trail, which crosses the southwest corner of the park, the only way to see the area surrounding the Hill Annex Mine today is by a bus tour that circles the open pit and includes three stops; on the second stop you can disembark to view overlooks, a boarding house, a miner's home, and equipment.

Besides the dangers of mining and a rugged lifestyle, the miners faced a communications challenge. The thousands of workers on the Iron Range included more than 40 ethnic groups.

Six mines once operated in this valley, and the tour bus passes the Pillsbury Mine, owned by John Pillsbury of flour mill fame, with the Dunwoody Mine visible in the distance. Somewhere along the tour your guide may mention the two-billion-year-old Laurentian Divide near here, a high plateau from which water drains in three directions: toward Hudson Bay and the Arctic Ocean, through the Mississippi River to the Gulf of Mexico, and through the Great Lakes/St. Lawrence River to the Atlantic Ocean,.

First leased for mineral exploration in 1892, Hill Annex, named after railroad magnate James J. Hill, was not actively mined until 1913, when it began as an underground mine but quickly converted to strip mining. Then, it was seriously excavated, producing 63 million tons of iron ore before closing. At the height of its operations, Hill Annex was the sixth largest producer of iron ore in Minnesota.

Hill Annex Mine State Park showcases large-scale Iron Range equipment. (Photo: Doris Green)

Early in the tour the bus passes the site of the mine's "location," at one time home to more than a dozen mining families. But as the mine grew, the homes were moved in the 1960s to further expand the pit. Workers had the option of buying these homes at rock-bottom prices.

When the mine closed, the pit was a half-mile wide at its widest point and had reached a depth of 500 feet. Groundwater seeps into the pit and water also enters from an underground river. The cold water in the pit is home to perch and even a few northern and walleye. Reclamation has brought wildlife back to the area, including the peregrine falcon, wolf, deer, fox, and an occasional bear.

Beyond the pit, the rolling landscape is also shaped largely by humans; many hills are former mine dumps. Aside from the ridge marking the Continental Divide, there's little natural ground left on the Mesabi Range.

The tour route provides many reminders that the Hill Annex pit did not always look like a wildlife preserve. You can get close to a few shovels and the big trucks (since replaced by even bigger trucks) ubiquitous on the Mesabi and see a wooden water tower. Watering trucks kept the dust down in the pit and on the roads.

Mine workers generally wore steel-toed boots and bib overalls and represented many different trades, such as mechanics, carpenters, and truck drivers. The highest paid was the shovel operator. While mining was generally a male domain, in World War II, fully half of Hill Mine's workers were women.

Prior to the war, each miner was issued a brass identification tag when he was hired. At the beginning of his shift, he would hand the tag to the "tool checker" and receive in exchange the tools he needed that day. At the end of the shift, he turned in his tools and retrieved his brass tag. Come payday, the miner would present his tag to receive his pay. You occasionally see these round brass tags—usually no more than two inches in diameter with a hole near the edge—in Minnesota antique shops.

The Hill Annex workers mined non-magnetic hematite, as opposed to magnetite, which is separated from other elements magnetically and formed into taconite pellets. At Hill Annex, miners extracted the iron by first crushing the ore and then pouring it into a "heavy media" or pudding including water, steel, scrap iron, and other materials. The iron sank and the material that floated to the surface was discarded. The process was repeated in additional slurries where the heavier iron again settled out.

Presently, Minnesota primarily provides taconite to the nation's steel mills. But as the resources of magnetic ore shrink and technology improves, the scenario may change. It's conceivable that in the future, industry may return to the lean ore remaining in the Hill Annex Mine and in the dumps surrounding it.

In addition to the ore, the historic mining site features a wealth of a different sort. Researchers have found sharks' teeth, giant coiled ammonites, and other fossils revealing the area's prehistory 86 million years ago when the Iron Range was covered by an inland sea. John Westgaard of the Science Museum of Minnesota in St. Paul and volunteer workers with the Hill Annex Paleontology Project also found a one-and-a-half-inch dinosaur claw bone at the park.

Directions:	From the Twin Cities, follow U.S. 169 north to Calumet. Then follow the signs to the park.
Seasons/Hours:	Open Friday and Saturday, 9 a.m. to 5 p.m., from Memorial Day weekend through Labor Day.
Length:	Tours last one and one-half hours.
Precautions:	Wear walking shoes if you plan to get off the bus to explore the replica mine buildings.
Amenities:	The visitor center offers exhibits, videos about mining, restrooms.
Information:	Hill Annex Mine State Park, Hill Annex Mine State Park C/O Scenic State Park, 56956 Scenic Highway 7, Bigfork, MN 56628. Phone: 218-247-7215 or 218-743-3362. Website: www.dnr.state.mn.us/state_parks/index.html and use the Park Finder.

HUPMOBILE TO GREYHOUND BUS

As World War I broke across Europe, workers in U.S. cities sometimes rode trolleys or jitney buses to their jobs. The miners on the Mesabi Range who did not live at a mine location did the same. A Hill Annex trolley ran along the line of pit mines from 1912 to 1927 when it went bankrupt. Beginning in 1914, a bus service that started with a seven-passenger Hupmobile carried miners from Hibbing two miles south to the nearby town of Alice, which later became a part of Hibbing.

Hupmobile? The Hupp Motor Car Company operated from 1909 through 1939, when it became a victim of the Great Depression. The miners' bus service began when Hibbing car dealer Carl Wickman used an unsold 1914 Hupmobile to begin the transit service with Andrew "Bus Andy" Anderson.

Their success led to more partnerships and mergers, and eventually the creation of the Greyhound Bus Lines. This creation story is told at the Greyhound Bus Museum in Hibbing through a video, exhibits, and many historic buses including a 1914 Hupmobile. The museum is open from mid-May through September.

Website: www.exploreminnesota.com/index.aspx and search for Greyhound Bus Museum.

HULL-RUST-MAHONING MINE OVERLOOK
ST. LOUIS COUNTY

The mine that grew—engulfing a town in the process—began shipping ore in 1895. Miners have excavated more than 800 million tons of iron ore from the Hull-Rust-Mahoning Mine in more than a century of operation. Named a National Historic Landmark in 1966, the world's biggest open pit iron ore mine is more than three miles long and two miles wide. It grew as many separate mines coalesced into the "Grand Canyon of the North."

Frank Hibbing held the first lease to mine ore on the site in 1891 and by 1893 he had founded the town that bears his name, and the first railroad had arrived. Ore was carried over the Duluth, Mississippi River and Northern Logging Railroad to Swan River and there transferred to the Duluth and Winnipeg line, which took the ore to the docks at Superior, Wisconsin, for shipment to the nation's steel mills. The original Mahoning Mine was the first open pit mine on the Mesabi Range, and soon strip mining replaced underground mining everywhere on the Mesabi.

The extraordinary view of the "Grand Canyon of the North" from the new Hull-Rust-Mahoning Mine Overlook. (Photo: Iron Range Tourism Bureau)

Ownership of the mine changed several times in its early years, and in 1899 James J. Hill of the Great Northern Railway bought Mahoning Mine along with the logging railroad and a tract of land. After Andrew Carnegie and Judge Gary organized U.S. Steel in 1901, its Oliver Iron Mining division took over operations at the Hull Rust pit. Today called Hibbing Taconite (or HibTac), the mine is owned by ArcelorMittal, Cleveland-Cliffs Inc., and U.S. Steel.

As the mine grew, it threatened the town of Hibbing itself, and from 1919 to 1921 about 200 buildings in what was then the heart of Old Hibbing were moved two miles south, to the site of the mining town, Alice Location. In 1935 the mine removed several more buildings close to the old town site.

Fast forward to 2018. The mine took over an observation area north of Hibbing that had attracted thousands of visitors annually. After being closed for the 2018 season, a new observation area opened in 2019 on a mine dump hill about a mile east of the old observation site.

Visitors can again see giant drills, dinosaur-like shovels as big as derricks, and a parade of monstrous trucks snaking slowly in and out of the pit. They are heading to the HibTac plant on the far horizon, where the ore is processed into taconite pellets for shipment to the steel mills of Indiana and elsewhere.

The new observation area features several observation stations constructed from Fronterra box culverts. The visitor center, 120-ton mine truck, and other equipment from the former observation site all moved to the new hilltop location. Standing inside of one of the huge shovels or in front of a giant mine truck helps visitors comprehend the magnitude of these mining operations.

This massive pit—and the miners who worked it—helped the nation develop and win two world wars. During peak production in the 1940s, almost one quarter of all iron ore mined in the United States came from this pit. Looking out over the mammoth hole, you can almost grasp its importance to Minnesota and the nation.

Directions:	From U.S. 169 in Hibbing, watch for signs to the Hull-Rust-Mahoning Mine overlook, located beyond the Greyhound Bus Museum in North Hibbing.
Seasons/Hours:	Relocated in 2019, the observation area is open daily from mid-May through September.
Length:	The world's largest open-pit iron mine is more than three miles long and two miles wide.
Precautions:	N/A
Amenities:	Hibbing offers plenty of dining, lodging, and recreation opportunities, website: http://www.hibbing.org/.
Information:	Iron Range Tourism Bureau, 111 Station 44 Rd., Eveleth, MN 55734. Phone: 218.749.8161. Website: https://ironrange.org/.

MINNESOTA MUSEUM OF MINING
ST. LOUIS COUNTY

The most fascinating underground attractions at the Minnesota Museum of Mining actually stand above ground: The steam and electric shovels, iron ore train, mine pump, and mineral samples will thrill any excavator, train enthusiast, or budding geologist. Granted, there is a replica of an underground drift that you descend into like a bomb shelter or small basement. But the multitude of informative artifacts positioned in the yard and buildings fronting the castle demand more attention.

Castle? The stone monument, complete with turrets, dates to the Great Depression of the 1930s. The Works Progress Administration (WPA) project originally formed part of a larger, community athletic complex that was converted to a museum more than 60 years ago. It's an appropriate venue for things historic, particularly anything related to stone, minerals, and mining.

Exhibits in the castle and other buildings in the 15-acre city park provide step-by-step guidelines for exploring, plumbing, drilling, blasting, ore washing, and other mining processes. A map depicts worldwide mineral deposits and another display describes ore transportation by rail and lake. There's a model of a taconite plant and a mining town, featuring mine office, blacksmith shop, newspaper office, school, and other replicas. Early firefighting equipment, railroad diorama, household furniture, and industrial and farm equipment link visitors with all aspects of life on the Iron Range.

The outdoor displays provide selfie opportunities galore:

- A 1904 100-ton Duluth, Missabe and Iron Range locomotive is "pulling" ore cars and a caboose.

- A Prescott steam-driven water pump kept the old Monroe Mine shaft dry and supplied Chisholm with a water source.

- A 1910 Atlantic Steam Shovel—a prototype of those used to dig the Panama Canal—is the only remaining example in existence.

- A truck park features a 35-ton Euclid dump truck of the type used in the pit mines during the post-World War II years.

- A 127-ton Wabco Electrohaul dump truck is now small in comparison to the trucks used in today's pit mines, but it demonstrates one stage in the ever-increasing size of the massive mining trucks.

- A Finnish log cabin is complete with sauna.

Since 2014, the nonprofit Igneous Metal Arts (http://igneousmetalarts.org/) of Minneapolis has held an annual Iron Pour at the Minnesota Museum of Mining. Old cast iron radiators are melted in a coke-fired blast furnace, then poured into molds to create cast iron art. Visitors can take part by scratching or carving designs in a block of sand that is poured with molten iron, creating a unique iron sculpture.

Directions:	From U.S. 169 northbound at Chisholm, exit on State Highway 73. (Head for the water tower and the castle.) Drive 0.75 mile and turn left at the water tower.
Seasons/Hours:	Open from June to early September, 9 a.m. to 5 p.m., Monday through Saturday, and 1 to 5 p.m. Sunday.
Length:	N/A
Precautions:	N/A
Amenities:	Chisholm offers plenty of dining, lodging, and recreation opportunities, see Chamber of Commerce website: https://chisholmchamber.com/.
Information:	Minnesota Museum of Mining, 701 W Lake St, Chisholm, MN 55719. Phone: 218-254-5543. Website: https://mnmuseumofmining.org/ or www.exploreminnesota.com and search for Minnesota Museum of Mining.

BRUCE HEADFRAME
ST. LOUIS COUNTY

The Bruce Headframe guards the eastern approach to Chisholm, rearing up suddenly on the north side of U.S. Highway 169. For a closer look, walk a half mile along the Mesabi Trail bordering the highway. The last on the Mesabi Range, the Bruce Headframe once held the sheave wheels for cables attached to cages and skips that lowered and lifted men and ore up a 300-foot shaft (now filled in).

Built by the International Harvester Company in 1926, the headframe was part of the mine complex that also included an engine house, dry house where workers changed, and a dozen two-story homes for workers. A sintering plant combined iron ore dust and other fine materials into a product usable by a blast furnace. The mine closed in 1948.

In 1978 the Bruce Mine Headframe was added to the National Register of Historic Places. A marker later erected at the site by the Iron Trail Convention & Visitors Bureau reads:

> Bruce Mine Headframe—This underground mine headframe hoisted low grade ore 300 feet to the surface. It is the last standing headframe on the Mesabi Range. In July, 1927, Nick Bosanich was reported to have died in a rock slide in the mine. Forty-six hours later he was found alive in a 10-foot-square room. His first request was for a cigarette.

More recently, the nonprofit Chisholm Beautification Association took the lead in developing Bruce Mine Park, working with the City of Chisholm, Iron Range Resources and Rehabilitation Board, St. Louis County, and others to create the 34-acre park at the headframe site. The association's mission is to "Develop a memorial park to recognize the sacrifice, dedication, and loyalty to the United States of America, by the thousands of immigrant miners and their descendants who developed the great Mesabi Iron Range."

The Chisholm Beautification Association held a groundbreaking ceremony at the site in 2018. Initial plans are for the construction of a road to the site, parking area, and brush-clearing around the headframe. Later development phases would involve illuminating the headframe and adding a lookout tower, an interpretative walk, and restrooms. In the course of project research, the association learned that a low-impact development would be preferable, in the event that mining might resume on the site in the future. The Chisholm Beautification Association is seeking additional supporting grants and donations.

Directions:	Visible from U.S. 169 at Chisholm, accessible from the Mesabi Trail.
Seasons/Hours:	Open year-round. Free.
Length:	N/A
Precautions:	N/A
Amenities:	Chisholm offers plenty of dining, lodging, and recreation opportunities.
Information:	Chisholm Beautification Association. Website: www.chisholmbeautification.org (See 2018 Projects).

MINNESOTA DISCOVERY CENTER
ST. LOUIS COUNTY

Previously called the Ironworld Discovery Center, the Minnesota Discovery Center is not to be confused with the Minnesota Museum of Mining, also at Chisholm. Both offer hands-on experiences and immersions into the Iron Range past.

You'll know you've reached Chisholm when the Iron Man statue comes into view. Officially titled "The Emergence of Man Through Steel," the 36-foot brass and copper miner stands atop a 49-foot steel structure created by sculptor Jack E. Anderson in 1987.

At the Minnesota Discovery Center, visitors climb aboard a 1928 trolley to ride to the recreated Glen Location. Plastered inside with photos of the past, the trolley carries visitors to another era. Along the way to the mining camp, the water-filled, two-mile-long pit of the Glen Mine offers visitors broad views of the rugged gorge. Other sights include an 1890 engine house replica, an actual 1890 railroad turntable, and a wooden water tower dating to 1910.

When your trolley pulls up at the Wilpen Depot, you step out into another time. The historic mining camp is complete with bunkhouse, bright yellow miner's home, and various mining accoutrements, as well as the Wilpen Depot. Both the Depot and tiny miner's house—once home to 13 people—were moved to this site. The original Glen

Across the highway from the Minnesota Discovery Center, the Iron Man stands tall. (Photo: Doris Green)

Location had 50 homes, often built close to the edge of the pit, to shorten the miners' walk to work.

When you've explored the camp and the pit to your satisfaction, you simply return to the Depot and wait for the return trolley. During the summer season, the trolley stops at the Depot every half hour.

Back in the main Ironworld complex, you can take advantage of various encounters with history.

- Outdoor living history exhibits include a Northwoods cabin, a pioneer homestead, and a depiction what life was like for the early northern Scandinavian immigrants, the Sami.

- Permanent exhibits feature the "Hall of Geology," "Underground Mining," and "Blue Collar Battleground: The Iron Range Labor Story."

- A Civilian Conservation Corps (CCC) History Center contains an amazing array of vehicles and other artifacts related to the work of the CCC in Minnesota, which employed more than 84,000 young men in 100 camps between 1933 and 1941.

Paved walkways and a bubbling stream meander through the park.

Directions:	From the south, follow U.S. 53 north to U.S. 169. Follow 169 west to Chisholm. Ironworld is located off 169 at Chisholm.
Seasons/Hours:	Open year-round: From Memorial Day weekend through Labor Day weekend, Tuesday through Saturday, 10 a.m. to 5 p.m. and 1 to 5 p.m. Sunday. Closed on Sunday during the winter. Fee. Extended evening hours on Thursdays with free admission.
Length:	To be fair to yourself (and the interests of your traveling companions) allow the better part of a day to explore the Minnesota Discovery Center.
Precautions:	Wear good walking shoes to tour the outdoor exhibits and the Glen Mine location.
Amenities:	Restaurant, living history exhibits, Minnesota CCC History Museum, Iron Range Research Center Library, mini golf with a mining theme, playground equipment.
Information:	Minnesota Discovery Center, 1005 Discovery Dr., Chisholm, MN 55719. Phone: 218-254-7959. Website: http://www.mndiscoverycenter.com.

MOUNTAIN IRON HISTORIC OVERLOOK
ST. LOUIS COUNTY

The Mountain Iron Historic Overlook is not just another pretty view of a deep blue water-filled mine pit set amidst iron-ore-red hills. This National Historic Landmark provides a view of the site where iron ore was first discovered on the Mesabi Range. Captain James Nichols made the discovery, while working for Cassius and Leonidas Merritt in 1890. The precise spot on the far slope of an open pit now lies on land owned by U.S. Steel's Minntac plant. After Nichols' discovery, a shaft was sunk and the first ore shipped to Duluth in 1892.

Behind the viewing station, a small park features mining equipment and a steam-powered 1910 Baldwin locomotive used to carry ore to Lake Superior ships. The photographer in your group will want to pose the entire family in front of the locomotive, as well as inside the Vulcan "Little Giant" steam shovel bucket. Another display in

This steam-powered locomotive on display in Mountain Iron once carried ore to Lake Superior. (Photo: Doris Green)

The "Little Giant" shovel bucket is a highlight of Mountain Iron's National Historic Landmark park. (Photo: Doris Green)

this outdoor museum maps the original site of the town of Mountain Iron, which moved south to allow mine pit expansion. Now a stop on the Mesabi Trail, the park today welcomes visitors from far and wide.

The 135-mile-long Mesabi Trail, running from the Mississippi River to the Boundary Waters, takes its name from a Native American word reportedly describing the glacier that once covered the area—a giant asleep upon the earth for many years. "Mesabi" was also the name of the winter constellation Orion.

On your way to the overlook, note the two-and-one-half-ton statue in front of the public library on Mountain Avenue. It stands as another tribute to Leonidas Merritt. Created by sculptor Robert Crump in 1940, the statue's base bears this inscription: "Leonidas Merritt, 1844 – 1926, Pioneer Prospector, Number One of the Seven Iron Men."

Directions:	From the Twin Cities, follow I-35 north to State Highway 33. Take 33 north to U.S. 53. Follow 53 north to U.S. 169. Follow 169 west to Mountain Iron. Exit onto County Road 102 and drive 0.75 miles north to Main Street. Turn left and drive two blocks to Mountain Avenue. The park is located at Mountain Avenue and Locomotive Street.
Seasons/Hours:	Open year-round.
Length:	N/A
Precautions:	N/A
Amenities:	Virginia offers restaurants, shopping, lodging, and good walking trails.
Information:	Information: Iron Range Tourism Bureau. Website: https://ironrange.org/attractions/locomotive-park/.

SEVEN IRON MEN

A legend on the Mesabi Range, seven sons and grandsons of Lewis H. and Hephzibah Merritt launched the mining industry and helped direct the development of early Duluth. Chief among them was Leonidas Merritt, gifted with an iron determination and energy to pursue the search for marketable ore and transport it to the steel mills of the eastern United States.

Like any legend, truth sometimes mixes with facts in the stories told about the Merritts. Though Leonidas did indeed have seven brothers, only three joined him in the pursuit of iron mining. Brothers Alfred, Cassius, and Andrus, plus nephews Wilbur, Bert, and John E. are usually counted as the iron tribe of seven. Two of Leonidas's brothers, Napoleon and Jerome, moved to Missouri, and Lucien became a minister and eventually began a pastorate at the Oneota Methodist Church in Duluth.

Lewis H. Merritt and son Napoleon came to Duluth from Ohio in 1855, with most of the rest of the family following in 1856. The family ran a hotel, and several sons began working in lumbering, surveying, and shipping. Leonidas and Cassius were timber cruisers, sampling and measuring a stand of woodland to estimate its potential for lumbering. From 1867 to 1868, Leonidas and Alfred worked as chainmen surveying for the Lake Superior and Mississippi Railway. A year later they constructed a schooner to ship goods, and a year after that they parted ways. Alfred continued the shipping business while Leonidas launched a company to buy and sell forest land. In 1886 Alfred was named St. Louis County Commissioner and Leonidas became Surveyor General of the Fifth Lumber District.

After survey team leader Captain James A. Nichols discovered iron ore in the Mesabi Range, the Merritts incorporated the Mountain Iron Company in 1890 and over the next two years came to own several mines and interests in several others in the Mesabi Range. Their big challenge: get the ore to the docks of Duluth. The Merritts created the Duluth, Missabe & Northern Railway Company to connect with an existing line. When that line faced financial difficulties, the American Steel Barge Company purchased a large interest of the Merritt's railway to construct a line all the way to Duluth, which was completed in 1893. The barge company was affiliated John D. Rockefeller.

(continued)

SEVEN IRON MEN (CONTINUED)

During the Panic of 1893, the Merritts faced growing financial problems. Leonidas negotiated a merger with Rockefeller, creating the Lake Superior Consolidated Mines Company. Rockefeller provided funds the Merritts needed to pay loans, which kept them solvent for a time. But as the depression worsened, they eventually sold all their Consolidated Mines stock to Rockefeller.

With huge success in sight, the Merritts lost all hope of wealth. Following subsequent lawsuits charging Rockefeller with misrepresenting the extent of his mining interests at the time of the merger, the Merritts eventually settled out of court, exonerating Rockefeller, who likely acted no more deceitfully than other industry leaders of the time.

Leonidas, Alfred, and Andrus then launched the American Exploration Company to search for copper and silver in the West, as well as Mexico and Canada. In 1913 Leonidas served as Duluth's Commissioner of Public Utilities and in 1921 became City Commissioner of Finance. Appointed to the Minnesota Old Soldiers Home Board of Governors in 1920, he served until his death in 1926.

LEONIDAS OVERLOOK
ST. LOUIS COUNTY

In any season, the sapphire lake and vermilion cliffs glint in the slanting rays of a late afternoon sun. Leonidas Overlook proclaims nature's beauty—except that it doesn't. The overlook stands as the highest *manmade* point on the Mesabi Range and the ephemeral lake may be gone in a generation. Continual advances in science and technology mean that the mine dump of today may well be the mine site of tomorrow—if it becomes profitable to mine "lean" ore.

Created from "overburden" or surface material including soil, sand, and rock, the Leonidas Overlook offers a 15-mile view of the Thunderbird Mine, a modern combination of several historic mines. One of these, the Leonidas Mine, opened in 1908 and was for a time the deepest underground mine in the world, with a depth of 650 feet. The Leonidas Mine evolved into a pit mine and produced nearly 24 million tons of iron ore before closing in 1980. The overlook stands squarely above the old Hull-Nelson Mine, originally

A wider lens than the one used here is needed to capture the extent of the mine view at Leonidas Overlook. (Photo: Doris Green)

Vegetation crowds in on an old mine building at the Leonidas Overlook. (Photo: Doris Green)

called the Adams. From 1901 to 1978 the Hull-Nelson Mine produced 19 million tons of ore. Portions of the Gross-Nelson and Spruce mines are also visible. When Eveleth Taconite formed in the 1960s, all the mines visible here coalesced into the Thunderbird Mine.

Visitors can see remnants of old mining operations, including an old mine building and areas in various stages of re-growth, with grass and trees slowly covering sections of the red earth. A line of hills visible to the northeast marks the Laurentian Divide, separating waters that flow to the Hudson Bay from those draining to the Gulf of Mexico or heading to Lake Superior and the Atlantic Ocean.

A joint project of the city of Eveleth and the Minnesota Iron Range Resources and Rehabilitation Board, the development of the Leonidas Overlook is documented on a fading sign posted a little way down the road back toward the county highway. The sign depicts a 1990 view of the area, relays the story of reshaping a stockpile into an overlook, and lists other completed reclamation projects.

Directions:	From the Twin Cities, follow I-35 north to State Highway 33. Take Highway 33 north to U.S. 53. Follow 53 to Eveleth. The overlook is 1.0 mile west of Eveleth on County Highway 101. The rough road up to the overlook is on the left-hand side of 101.
Seasons/Hours:	Dawn to dusk year-round, however, snow and spring rains may lead to temporary closures of the gravel road leading up to the overlook. Free.
Length:	N/A
Precautions:	There are drop-offs; carefully supervise young children.
Amenities:	No facilities. Eveleth offers dining, lodging.
Information:	Iron Range Tourism Bureau, Station 44 Rd., Eveleth, MN 55734. Phone: 218-749-8161-111. Website: https://ironrange.org/attractions/leonidas-overlook/.

MINEVIEW IN THE SKY NO MORE (FOR NOW AT LEAST)
ST. LOUIS COUNTY

Until October 2015, the "King of the Load" stood guard at the entrance of the Mineview in the Sky in Virginia, Minnesota. Dwarfing all the vehicles that climbed the drive to the former observation area, the 240-ton, bright yellow mining truck made a lasting impression on visitors.

But no more. After 36 years of offering amazing views of the operating United Taconite Mine and the abandoned Rouchleau Mine pit, both the overlook and visitors center closed to make way for the growing taconite mine. Even U.S. Highway 53 was rerouted, including a new, 200-foot-tall bridge over the Rouchleau Mine lake completed in 2017.

Visitors can view the Rouchleau
Mine from this caged bridge in a
Virginia city park.
(Photo: Doris Green)

Visitors can now see the partially water-filled, mined-out pit from the new bridge and the Mesabi Trail, which runs along the east side of the bridge on Virginia's southern edge. One day they may have easier access to this view as Virginia has considered creating a new scenic overlook over the Rouchleau Mine lake.

The Rouchleau Mine, named after Louis Rouchleau of Duluth, and many of the other nearby mines—including the Lone Jack, Union, Moose, and Enterprise—began as underground shafts. Eventually their maze of tunnels met and intertwined underground. By the 1930s, however, mining here shifted to pit mining, partly because the ground was settling and somewhat unstable and partly because technology improvements made pit mining more feasible. The original shafts and tunnels were engulfed in one huge pit.

U.S. Steel and its predecessor companies have owned the Rouchleau Mine since 1893. It last produced iron ore in 1977.

To see a different side of the Rouchleau Mine, head into Virginia to Finntown at the end of Third Street near Kaleva Hall. A city park offers picnic tables, access to the Mesabi Trail, and a 50-foot caged safety bridge protruding over the pit. The modernistic deck entrance of blue steel pillars topped by a pair of triangular trusses frames the panorama.

Directions:	From the Twin Cities, follow I-35 north to State Highway 33. Take Highway 33 north to U.S. 53. Follow 53 to Virginia. Access the Mesabi Trail by taking the Second Avenue exit from Hwy 53 into Virginia. Turn right on Chestnut Street and park in front of the greenspace on the north side of the street. The paved trail crosses Chestnut here; walk south (right) to access the trail. Walk (or bike) 0.8 mile on the trail to the east side of the bridge. The Finntown overlook can be found at the east end of Third Street.
Seasons/Hours:	Open daily, dawn to dusk.
Length:	The pit is about 3.0 miles long and 0.5 mile wide.
Precautions:	Pedestrians and cyclists share the Mesabi Trail.
Amenities:	Virginia offers shopping, trails, dining, and lodging.
Information:	Virginia Area Chamber, P.O. Box 1072, Virginia, MN 55792. Phone: 218-741-2717. Or, visit the DNR website: https://ironrange.org.

LAKE ORE-BE-GONE
ST. LOUIS COUNTY

Nothing woebegone about Lake Ore-Be-Gone. Created from three former iron ore pit mines (Gilbert, Schley, and Pettit), Ore-Be-Gone lures water skiers, swimmers, fishers, and scuba divers who find submerged mining equipment—not to mention a helicopter, school bus, cars, plane, and a diverse fish population—in the 140-acre, 400-foot-deep lake.

Also the center of Minnesota's only off-highway recreation area, Lake Ore-Be-Gone buzzes in summer with the traffic of all types of off-road vehicles. Whatever their sport, many visitors opt to stay at Sherwood Forest Campground on the west side of the lake. Developed with help from the Iron Range Resources and Rehabilitation Board, the campground provides trail access to the Iron Range Off-Highway Vehicle Recreation Area and lies adjacent to the Mesabi Trail.

Ore-Be-Gone is not the only former pit mine site that has evolved into a scuba diving recreation area. Lake Mine Quarry in Biwabik and Stubler Mine 12 miles west of Virginia offer other Mesabi Range dive sites. To the southwest in the Cuyuna Range, divers may spot big northern pike, plastic skeletons, and mermaids from some of the 55 dive sites at 27 Crosby mine pits.

One of several pit lakes off U.S. 169 between Chisholm and Virginia. (Photo: Doris Green)

Drive almost anywhere through Minnesota's Iron Range and you will see many water-filled pit mines and reforested hills. Several of the new lakes, like those along U.S. Highway 169 between Chisholm and Virginia, feature gravel parking areas and boat launches. Thanks to all the reclamation efforts, the landscape appears almost as wild as when Leonidas Merritt (See SEVEN IRON MEN) hiked these woodlands.

Directions:	From Duluth, follow U.S. 53 north about 37 miles and turn right on State Highway 37. In Gilbert, turn right on Wisconsin Avenue and drive three blocks to Sherwood Forest Campground. (Watch for signs to the campground in Gilbert).
Seasons/Hours:	Open year-round.
Length:	N/A
Precautions:	There have been accidents in the pit lakes. If you scuba dive, take a dive buddy and follow safe diving practices.
Amenities:	Gilbert has a few good restaurants and back on U.S. 53, Eveleth offers several more plus lodging.
Information:	Sherwood Forest Campground, 101 W Wisconsin Ave E, Gilbert, MN 55741. Phone: 218-748-2221. Website: https://www.exploreminnesota.com; click on Places to Stay/Campgrounds.

LONGYEAR DRILL SITE
ST. LOUIS COUNTY

The timber frame points skyward like a triangular teepee. It doesn't look like much, but from this humble beginning grew the Longyear Exploration Company, which provides core-drilling services around the globe. The Longyear Company has searched for iron ore in Minnesota, Venezuela, and Peru, copper in Arizona and Rhodesia, and zinc and silver in Argentina and China. Its construction programs even included test borings for San Francisco's Golden Gate Bridge.

Edmund J. Longyear arrived on the Mesabi Range in 1890 as a fresh graduate of the Michigan Mining School at Houghton. He set up "housekeeping" in a one-room shack with his bride, Nevada, and built a steam-drilling rig a mile or so away with two partners.

Longyear sunk a diamond-bitted drill 1,293 feet into the earth, the first time anyone had used a diamond-bitted drill on the Range. The drill produced core samples that could

Doris Green at the Longyear Drill Site, surrounded by vegetation in 2018.
(Photo: Michael H. Knight)

be tested for mineral content. Although Longyear did not find evidence of usable ore, he did begin a new era in iron ore exploration.

The E.J. Longyear Company began offering its contract services to mine developers. From 1890 through 1911, Longyear drilled hundreds of samples along the 110-mile Mesabi Range.

In 1911 Longyear expanded the company, buying out his partners (one was his uncle) and establishing the Longyear Exploration Company. Over his career, Longyear directed the drilling of more than 7,000 test holes in Minnesota and far beyond its borders.

The original Longyear drilling site is down a quarter-mile trail that begins behind the information booth in the parking lot. In addition to the timber frame above the test site, the restored site features a drill that would have held the diamond bits, a steam boiler, a water pump to cool the bits, and a churn buck to drive down and later remove the drill casing.

In 1976 the Iron Range Historical Society reconstructed Longyear's first drill site north of Hoyt Lakes and installed authentic diamond drilling equipment. The State and National Historic Site is maintained by the City of Hoyt Lakes and the Hoyt Lakes Garden Club.

Directions:	From the Twin Cities, follow I-35 north to State Highway 33 near Cloquet. Take 33 north to U.S. 53 north to State Highway 37. Turn east on 37 and drive through Gilbert to Aurora. In Aurora, turn on County Highway 100 east to Hoyt Lakes. In Hoyt Lakes, turn left on County Highway 110. About 3.0 miles north of Hoyt Lakes, turn right on County Highway 666. Look for a parking lot on your right with an information booth on the far side.
Seasons/Hours:	Dawn to dusk, spring through fall. Free.
Length:	N/A
Precautions:	Limited access to individuals with disabilities. Quarter mile trail through the woods.
Amenities:	Hoyt Lakes offers dining, lodging.
Information:	Iron Range Historical Society, 5454 Grand Ave., McKinley, MN 55741. Phone: 218-749-3150. Website: www.ironrangehistoricalsociety.org.

SOUDAN UNDERGROUND MINE STATE PARK
ST. LOUIS COUNTY

When you first catch sight of the 90-foot headframe looming over the mineshaft and hear the roar of the hoist motor, you may have second thoughts about climbing into the "cage" and descending a half mile into the Soudan Underground Mine But don't let a moment's hesitation rob you of a unique down-under adventure in Minnesota's oldest and deepest underground mine.

Buy your tickets. Don your hardhat. And climb bravely into the cage when your turn comes. As it moves downward, you watch the openings of levels slip by. It may feel as if you're dropping at the speed of light, but actually it's little more than 10 miles per hour. And although the trip feels like a noisy elevator ride, it's not a straight descent. The shaft is at a 78-degree slant.

Unlike an elevator, the cage has wheels on one side that ride on tracks laid in the side of the shaft, which is lined with concrete and iron. Beneath the cage once hung a skip that carried up to six tons of ore. At the top, the cage is attached to hoist ropes that rise up the headframe and over the giant sheave wheel. From there, the ropes run into the nearby engine house, which contains the enormous hoist motor. Traditionally, an engine house was located a little distance from the shaft, to ensure that vibrations from the engine did not dislodge rocks in the mine.

The cage stops at the mine's 27th level, 2,341 feet beneath the surface. There are 50 miles of tunnels in the Soudan Mine and you begin your journey through them via a train, which runs three quarters mile into the tunnel network. The ride is both bumpy and breezy, as wind sails unrestricted through the vented nine-by-nine drift or horizontal passage.

The train stops below the Montana Ore "Stope" or void created by the extraction of ore. To hollow out a stope, the miners cut upward from the drift into a vein of ore, extracting both ore and debris. The Soudan is a hard mine; no timbers were needed to support the stopes as they were built. When the workers mined upward, they filled the floor with debris, raising its level. This process is called the cut-and-fill method of mining.

Three openings led into a stope: one for the transfer of ore, one with a pulley to raise equipment into the stope, and a manway with a ladder for the miners to climb. Today visitors enter the Montana Ore Stope via a spiral staircase installed in the manway.

At this point in the tour, the guide may turn on a drill, cautioning you to "cover your ears." The roar is deafening and the guide observes that the miners who once worked here never wore protective earmuffs or dust masks.

The early miners used a tugger that powered a scraper to scrape the loose chunks of ore into a deep hole or chute. The ore was loaded into a "Granby car" or ore car and transported to the shaft, where miners loaded the ore into the skip.

At left, the Soudan Mine headframe rises above the shaft; an elevator takes visitors a half mile down into the underground mine. (Photo: Doris Green)

After leaving the Soudan Mine crusher house, the iron ore was loaded into rail cars. (Photo: Doris Green)

No one has mined here since 1962. Although considerable iron remains underground, it's not currently feasible to extract it. During 80 years of mining operations, 15.5 million tons of ore were removed from the Soudan Mine.

Iron was discovered here during the short-lived 1865 Vermilion gold rush; however, an open pit mine was not established until 1882. The mine encompassed as many as seven pits. The boardwalk at the end of the parking lot leads visitors to the deepest pit. All of the Soudan pits tended to be deep, since the ore body headed almost straight downward into 2.7-billion-year-old metamorphic bedrock. The deep pits were dangerous. Rocks would frequently and unexpectedly fall from the walls, and the mine was soon converted to an underground operation.

While the ore on the Iron Range tends to run in horizontal layers, the ore bodies at the Soudan Mine run in almost vertical (78-degree) columns. The rock layers have been deformed by the movement of tectonic plates and volcanic eruptions as the North American continent developed.

The early underground miners worked by dim candlelight, often carrying a shift's supply of candles in a Prince Albert tobacco tin, curved to fit comfortably into a back pocket. It was not until 1918 that carbide (gas) lamps were introduced in the mines, and around 1940 wet-cell batteries came into use.

From the beginning of mining at Soudan, much work took place above ground, and several buildings remain. The park's ticket office is located in the old dry house, where the miners would hang their wet overalls to dry at the end of a shift. You also can tour the engine house and see the crusher house and the chutes that once carried crushed ore into waiting rail cars. The drill shop is also open for viewing.

In addition to the regular Underground Mine Tour, the park began offering geology-focused walking tours in September 2018. Instead of taking a train for an approximately three-quarter-mile ride through the underground passages, visitors can walk the route and see points of geologic and mining interest that train riders never see. Wearing headlamps on their hard hats, visitors examine rock formations, look for clues to ore deposits, and walk beneath a fault line.

In 2019 the park added regular scientific tours to its schedule, which also visit the 27th level of the mine. Science nerds can see the former underground physics laboratory that hosted the Main Injector Neutrino Oscillation Search (MINOS) to detect and study a fundamental particle, the neutrino, beamed to the mine from Fermilab near Chicago.

Tour participants also learn about current research involving deep water that bubbles up through the 2.7-billion–year-old rock into old drifts and crosscuts from bore holes drilled years ago by miners seeking rich iron ore. The deep waters bubbling up are a calcium, sodium, magnesium chloride brine about twice as salty as sea water. This brine contains ferrous iron and is anoxic (lacks any oxygen gas). It's also an environment inhabited by unusual bacteria and other single-celled microbes. When these anoxic waters reach the air in the mine they begin to adsorb oxygen and deposit formations similar to those found in caves: flowstones, stalagmites, soda straw stalactites, and rimstone dams. These formations are made of the iron oxides ferrihydrite and goethite with colors ranging from oranges and reds to black. Learning more about this metal precipitation could lead to potential biofuels.

When E. Calvin Alexander, Jr., Morse-Alumni Professor Emeritus with the Earth Sciences Department at the University of Minnesota, saw these formations, they reminded him of photos of the surface of Mars. While photos of Mars have shown gullies and other features that appeared to be the result of flowing water, scientists discounted this theory because liquid water could not exist on the very cold planet; however, the brine bubbling up in the Soudan Mine might be a model for conditions on Mars.

Some microorganisms in the brine have anti-fungal properties and might one day lead to a biocontrol for white nose syndrome, thanks to research led by Christine Salomon, associate professor at the University of Minnesota's Center for Drug Design (see WHITE NOSE SYNDROME). WNS has been confirmed at the park.

Directions:	From the Twin Cities, follow I-35 north to State Highway 33 near Cloquet. Take 33 north to U.S. 53 north through Virginia. Turn northeast on State Highway 169 and drive to Soudan, Minnesota. Then follow the signs to the park, located about three blocks off of 169. The park is 225 miles from the Twin Cities.
Seasons/Hours:	Park is open daily from 9:30 a.m. to 6 p.m. Underground Mine Tours run regularly each day, from Memorial Day weekend through September and weekends in October. The rest of the year, call for group or educational tours.
Length:	The Underground Mine Tour and Geology Tour traverse about 0.75 mile of passages. Allow an hour and a half for the Underground Mine Tour.
Precautions:	Wear good walking shoes and bring a jacket or sweatshirt since the mine is 51 degrees year-round. To help prevent the spread of white nose syndrome, visitors walk across special mats designed to remove fungal spores from footwear. Visitors are instructed not to visit other caves or mines with any clothing, footwear or gear they have used in areas where the fungus is present without cleaning it properly first. There are two spiral staircases on the Underground Mine Tour, but all tours are accessible. Call for details. No bags or strollers are allowed underground.
Amenities:	Visitor center, nature store, restrooms, picnicking, hiking, fishing, camping, geocaching, and snowmobiling. Self-guided audio surface tour of the mining operations available; check out MP3 players from the visitor center at no charge.
Information:	Lake Vermilion-Soudan Underground Mine State Park, P.O. Box 335, Soudan, MN 55782. Phone: 218-300-7000. Website: www.dnr.state.mn.us/state_parks/index.html and use the Park Finder.

PIONEER MINE
ST. LOUIS COUNTY

North of downtown Ely, the Pioneer Mine headframe and water tower provide a focal point on the horizon, with Miners Lake in the background. Pioneer Mine opened in 1888, first shipped ore in 1889, and closed in 1967; its history is the history of Ely.

Placed on the National Register of Historic Places in 1978, the mine is now owned by the city and administered by the nonprofit Ely Greenstone Public Art Committee as the Ely Arts & Heritage Center. The Shaft House, which contains mining artifacts and displays, is open to the public Tuesday and Friday afternoons in the summer and by appointment. Not far away, down the stairway toward the lake, the Dry House is a seasonal, rentable event venue and used for meetings, art classes, and exhibits.

Miners Lake, a designated trout lake, formed from a combination of surface mining endeavors, earth settling, and collapsing underground mine passages. Some ore here is quite soft and mud-like when wet, contributing to the dangers, difficulties, and expense of underground mining. According to the text of a city marker: "There was constant danger that water would soak the ground above, break through and run into the mine, sometimes to a depth of 200 feet, killing all those not fast enough to escape."

Considerable deposits of rich hematite still remain; however, open pit mining is generally less costly and more profitable than the underground mining required at the Pioneer and nearby mines. All told, five mines once operated here: the Zenith, Savoy, North Chandler, and South Chandler, in addition to the Pioneer.

Directions:	From Duluth, take U.S. 53 north and turn northeast on State Highway 169 (169 changes from a U.S. highway to a state highway at Virginia). Drive 45 miles to Ely. From downtown Ely, follow County Highway 21 (Central Avenue) to the north side of Miner's Lake. You also can reach the park via the Trezona Trail, a 5-mile recreational trail that circumnavigates the lake.
Seasons/Hours:	Park is open daily, dawn to dusk; the Shaft House is open Tuesdays from 2-5 p.m. and Fridays from 3-6 p.m. and by appointment 218-365-2841.
Length:	N/A
Precautions:	Boaters should watch for timbers and other relics lurking in Miners Lake.
Amenities:	The Trezona Trail heads toward the shopping, dining, and lodging opportunities of Ely.
Information:	Greenstone Public Art, 401 North Pioneer Rd., Ely, MN 55731 US. Phone: 218-235-1721. Website: https://elygreenstone.org/. Dry House rental information: https://www.ely.mn.us/?SEC=673EC7CC-EB1E-4D02-B568-7B8366526579.

Placed on the National Register of Historic Places in 1978, the Pioneer Mine is now the Ely Arts & Heritage Center. (Photo: Doris Green)

VOYAGEURS NATIONAL PARK
KOOCHICHING AND ST. LOUIS COUNTIES

To get the feel of the late 18th and early 19th century voyageur experience while you explore an old mining site, sign up for the Grand Boat Tour at the Rainy Lake Visitor Center. As the cruise boat threads its way among the islands, close to the 2.7 billion-year-old rocky shorelines, you may feel you've dressed wrong for the occasion: a voyageur's moccasins, a bright sash at your waist, and a large hat to keep out wind and rain would be more suitable than your t-shirt and sandals.

Getting to one of the park's gold mines requires a boat trip from spring through fall; you can take the park's cruise service, your own boat, or a rented canoe, motorboat, or houseboat. During winter, you can ski, snowshoe, or snowmobile to the mines. In any case, you will likely notice wildlife and wooded wilderness terrain on your trip.

Established in 1975, Voyageurs National Park contains more than 30 lakes, as well as multiple bogs, marshes, and beaver ponds. You can access these waters from entry points at three visitor centers, where you can leave your car and switch to water transportation.

Less known than the 1849 California Gold Rush, the Rainy Lake Gold Rush began in 1893, when George W. Davis discovered gold in a vein of quartz at Little American Island.

There are still veins of quartz containing gold in a belt of rock schists extending from Little American Island to about 50 miles west and more than 120 miles east into Canada. This ancient rock formed more than two billion years ago along the now-inactive Rainy Lake-Seine River fault. When hot magma was thrust to the surface and then cooled, it trapped gold and other minerals with lower melting points into cavities in the quartz. Part of the Canadian Shield, the greenstone in the park represents some of the oldest exposed rock on the globe.

Davis' discovery set off a gold boom that lasted only until 1898. But during those five brief years, men—and a few women and children—flocked to the region. Shafts were sunk, tramways and crushing mills developed, and towns created. Spurred by the Panic of 1893, which left many people out of work, and by the lure of wealth and adventure, they headed north. Many hoped to work in the mines; others planned to offer whiskey, poker, and other amenities to the miners in the new boomtowns.

The Little American Mine was originally owned by Davis and Charles Moore, who had grubstaked Davis and earlier developed a mine at Lake of the Woods. But developing a mine was a costly undertaking, so Davis and Moore sold their claim to Jeff Hildreth, who brought in additional investors and dug an adit, or horizontal tunnel, into the island. He then built a mill at Rainy Lake City to crush the quartz and sank a vertical shaft. At the 35-foot level of this shaft, miners hollowed out the first stope, or chamber, hoisting the ore to the surface via a large pulley. Eventually, miners extended a second shaft to a depth of more than 200 feet.

Rainy Lake City developed quickly to support the exploration and mining endeavors. Incorporated in 1894, the city soon numbered several hundred people and boasted a school, bank, general store, hotels, restaurants, a newspaper, hardware store, and multiple saloons.

Yet, the Little American Mine turned a profit only briefly, overcome by a series of obstacles and under-capitalization. It changed ownership rapidly, as one investor after another sunk funds into the development of the Little American. One year, a worker let $10,000 worth of gold slip into the Rainy Lake. Another year, a loading dock collapsed, sending an entire season's work of ore into the lake. Finally in 1898, the sheriff seized the mine, and later attempts to re-open it ended in failure as the Klondike Gold Rush attracted miners and mining capital to the Alaskan frontier.

Placed on the National Register of Historic Places in 1975, the Gold Mine Historic District is one of 16 historic properties in the park. All told, there were about a dozen gold mines here, including the Lyle Mine north of Dryweed Island, the Big American

Mine on Big American Island, the Bushyhead Mine on Bushyhead Island, and the Soldier Mine on Dryweed Island.

An easy way to access these sites is to board a boat at the Rainy Lake Visitor Center and take the Grand Tour. The small cruise boat passes islands where buildings mark the site of an old mining village, and you can see the entrance to the adit of Bushyhead Mine on Bushyhead Island. This tunnel extends halfway through the island to a vertical shaft, now sand-filled for safety. The tour boat also motors past Powder Island, where the early miners stored blasting powder at some distance from their mining operations—as a precaution against lightning strikes.

When the Grand Tour reaches Little American Island, you can disembark and stroll along an accessible, quarter-mile trail leading to historic mine shafts and equipment. Signage along the way explains early mining operations.

Back on the boat, the ranger-led tour continues, You may see treeless islands in Canadian waters that are rookeries for gulls and cormorants. You will hear their cacophony long before you can clearly see these birds. You may also see goshawks, herring gulls, and perhaps a loon with a baby on its back. Bring your binoculars and look for "golf ball" bald eagle heads in the treetops.

Other park boat tours take you close to the Ellsworth Rock Gardens and to the Kettle Falls Hotel, accessible only by water. The park also offers voyageur canoe programs and naturalist-led programs on the Northern Lights and area wildlife.

Directions:	From International Falls, follow State Highway 11 east for 11 miles to reach the Rainy Lake Visitor Center.
Seasons/Hours:	Year-round. The Grand Tour is offered from spring through fall.
Length:	N/A
Precautions:	Make boat tour reservations early; they do sometimes sell out. If you travel by private motorboat, houseboat, or canoe, heed local weather forecasts. Storms can come up quickly. If you travel in winter, beware of slush and unsafe ice conditions, as well as blizzards.
Amenities:	Rainy Lake Visitor Center is open all year. Ash River and Kabetogama Lake Visitor Centers are open during the summer. All offer books, maps, navigational charts, and exhibits. Summer naturalist programs provide information on the wildlife and history of the area.
Information:	Rainy Lake Visitor Center, Voyageurs National Park, phone: 218-286-5258. Website: www.nps.gov/voya.

ROBINSON'S ICE CAVE, BANNING STATE PARK
PINE COUNTY

Robinson's Ice Cave is Minnesota's closest approach to a true ice cave. A towering cave opening, sealed with a ladder-like gate to keep people out but allowing bats to enter, presents itself in Banning State Park, near Sandstone, Minnesota. Following a fracture in the Precambrian Hinckley Sandstone for a distance of 200 feet, this cave serves as a DNR-protected bat hibernaculum. A large boulder, wedged where it fell, overhangs the passage. Captured on an early postcard, the boulder remains in the same position today. Legend has it that the cave extends under the Kettle River—something that we hear frequently about riverside caves.

But Banning State Park has even more to offer. Between 1892 and 1912 these woodlands reverberated with the sounds of blasting and cutting, as well as the thundering of the Short Line of the St. Paul-Duluth Railroad. Hundreds of quarrymen worked drilling, moving, and shaping the pink sandstone into building stones, paving stones, curbs, and bridge abutments. The quarrying boom ended as the supply of good stone diminished, along with the demand for sandstone as a building material. Increasingly, structural steel and concrete replaced the natural sandstone.

Although part of Banning State Park, Robinson's Ice Cave is more readily accessed from Robinson City Park on the east edge of the town of Sandstone. Park and continue on foot, following the broad, easy Banning Quarry Trail running north, past building foundations and the half-mile long quarry face, into the state park, under the railroad trestle. This easy trail ends at the north end of the quarry (one-fifth mile north of the trestle) where you'll find a small cave in the reddish sandstone, not to be confused with the ice cave. Beyond the quarry, a faint footpath follows the riverbanks over occasionally wet ground; hike one third of a mile north of the trestle before turning left into the woods along an unmarked foot path and going up to the ice cave in the cliff face. The cave entrance is covered with a large steel gate but you can look through the horizontal bars and see the cave quite well, including the immense boulder wedged in the passage and the bat hibernaculum signage. You can sometimes see bats leaving the cave at dusk in summer.

As defined by Alpine scientists in Europe, ice caves contain perennial (year-round) ice. In Minnesota, with its modest elevations and mid-latitude position, such conditions do not exist, and only seasonal ice can be found in our version of an ice cave. This precludes the notion that the ice in this cave is a relic of the last Ice Age, as well as expecting to extract a long-term paleoclimate record from the ice, as has been done elsewhere.

There are two kinds of cave ice: nivation ice and regelation ice. Nivation ice forms by snow blowing into a cave entrance and morphing into ice layers. Regelation ice forms by spring meltwater dripping into a cave and refreezing in the cold air trap created by the cave. The ice in Robinson's Ice Cave is strictly the latter kind. In the spring of each year,

Bat gate at the entrance to Robinson's Ice Cave. (Photo: Greg Brick)

A large boulder overhangs the passage in Robinson's Ice Cave. (Postcard from the Gordon Smith Collection)

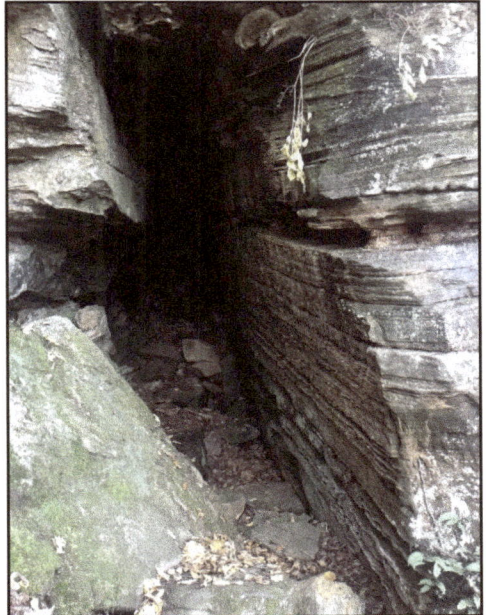

Small cave downhill from Robinson's Ice Cave. (Photo: Greg Brick)

club-shaped ice stalagmites, some as tall as a person, grow up from the floor. Hardly comparable to the elephantine ice formations of European caves, but it's what we have.

For another cave experience, there's a "consolation cave" immediately downhill from the ice cave, formed in a vertical joint where the cliff is pulling away. About 30 feet long, the cave's walls display crinkly layered bedding and are dotted with "tafoni," pockets caused by weathering of the sandstone.

Directions:	From the Twin Cities, follow I-35 north. Use Exit 195 for the main state park entrance, but Exit 191 for the caves. From Exit 191, drive State Highway 30 east into town of Sandstone, then State Highway 23 south to Robinson City Park on the left side, just before the bridge over the Kettle River. The city park adjoins the state park on the latter's southern boundary. GPS coordinates for Robinson's Ice Cave are UTMs 510677 East, 5109830 North.
Seasons/Hours:	Year-round, 8 a.m. to 10 p.m. During the winter, however, many campsites, water, and restrooms are closed and Robinson's Ice Cave is inaccessible. Park fees.
Length:	The Banning Quarry Trail is the shortest way to the ice cave, less than one mile. Hiking south to the cave from the state park proper is a greater distance, but allows you to see additional features including the Hell's Gate Rapids of the Kettle River, along with the potholes that gave the river its name.
Precautions:	Trails vary in difficulty. The Banning Quarry Trail is moderately easy, encompassing small hills and ravines. The trail to Robinson's Ice Cave is rugged and not recommended for young children or people with limited mobility. The final approach to the cave involves stepping among boulders.
Amenities:	Banning State Park offers a gift shop, camper cabin, campgrounds, and canoe campsites, boat landings, picnic areas, overlooks, recreational trails, and occasional naturalist programs. During winter there are cross country skiing trails (not groomed), as well as snowshoeing, and snowmobiling. Robinson's Park offers bouldering, ice climbing, and access to the Kettle River.
Information:	Banning State Park, P.O. Box 643, Sandstone, MN 55072. Phone: 320-245-2668. Website: www.dnr.state.mn.us/state_parks/index.html and use the Park Finder.

ELY'S PEAK TUNNEL
ST. LOUIS COUNTY

Entering Ely's Peak tunnel, many visitors grow silent, peering into the gloom ahead and treading with care on the former railroad bed. A stiff breeze wraps round them and quickly carries away any words they do speak. Just as the light from the tunnel entrance behind them fails, they notice a glow ahead. Although the 469-foot tunnel curves, flashlights aren't necessary to explore it. Standing in the middle of the tunnel, visitors can clearly see both entrances. But you'll greatly enhance your tunnel experience if you do have a light, as it will help you to examine details in the rock walls. See if you can detect the layering in the basalt lava flows. Can you tell which way the lava was flowing—toward or away from what is now Lake Superior (where the magma chamber was located)?

Dug out of basalt, a dark volcanic rock, the tunnel was constructed about 350 feet above Lake Superior in the late 19th century. It was used by the Duluth Winnipeg and Pacific Railroad until the company abandoned the line in the 1970s. Now owned by the city of Duluth, the old railroad bed is a recreational trail, linking to the Willard Munger State Trail, which begins (or ends) in Duluth. Named for the Minnesota state representative who devoted his career to protecting the environment, the Willard Munger State

An abandoned railway tunnel bores through the heart of old lava flows. (Photo: Greg Brick)

Trail is one of the longest paved trails in the United States. It and several other state trails also travel through a number of short, concrete tunnels underneath highways, but none compare in magnitude or mountainous beauty to the Ely's Peak tunnel.

Ely's Peak is not located anywhere near the town of Ely, but on the outskirts of Duluth. If you travel to the tunnel by the Munger or the Duluth Winnipeg and Pacific Trail, you can hike or cycle. Your dog is welcome, too, although it must be on a leash. In winter, you can use snowshoes, cross-country skis, or a snowmobile. The tunnel is some 12 miles south of the Willard Munger Inn, where you can find a trail map, lodging, refreshments, and bicycles to rent.

You also can access the old railroad trail from a parking area on Beck's Road. This route involves a challenging 15-minute hike up a steep, rocky trail.

The Duluth Winnipeg and Pacific Trail runs through Tunnel Bluff, one of two cliffs (the other is Northwestern Bluff) on Ely's Peak, near the Spirit Mountain area southwest of Duluth. The peak is named for Edmund Ely, an early Presbyterian missionary and pioneer who established the first school in St. Louis County in 1834. Today, rock climbers ascend several routes up these bluffs; most routes involve climbing via numerous vertical cracks in the rock faces. You do not, however, need to be a rock climber to reach the top of the peak, which offers fantastic views of the valley below.

Directions:	The trailhead is at Willard Munger Inn, located in Duluth across Grand Avenue from the Lake Superior Zoo. You also can access the tunnel from the Beck's Road Willard Munger State Trail parking area. Take Grand Avenue south to Beck's Road (County Highway 3) and turn right. Then, turn right again on 123rd Avenue West, which ends at the parking area. (If you miss this turn, you'll intersect I-35 in 2.5 miles.) From the parking area, go to the Munger Trail. Head up the trail to your right, past a house and driveway. In approximately 0.1 miles take the trail to the left, which leads across a field. Cross the railroad tracks—but watch for trains since these tracks are in use today. The area around this crossing is often muddy. Follow the red dirt and gravel trail into a predominantly birch woods. Here, the trail climbs upward, becoming rocky and difficult, but soon intersects with the old railroad bed. You can see the tunnel directly to your right. If, instead of heading into the tunnel, you cross the old railroad bed, the trail continues to the top of the peak, offering a magnificent panorama.
Seasons/Hours:	Year-round. Free.
Length:	The curved tunnel is 469 feet long.
Precautions:	Getting to the tunnel involves either a 12-mile easy hike, bike, snowmobile, or cross-country ski trip, or it requires a challenging seven- to 10-minute hike up a steep, rocky trail.
Amenities:	The Willard Munger Inn offers single and double rooms and studios, some with whirlpools and fireplaces. In winter, the Inn has special ski packages; the rest of the year if offers free bicycle use with a room and adventure bike tours including shuttle service.
Information:	Willard Munger Inn, 7408 Grand Ave., Duluth, MN 55807. Phone: 800-982-2453 or 218-624-4814. Website: www.mungerinn.com .

NOCTURNAL BUILDING, LAKE SUPERIOR ZOO
ST. LOUIS COUNTY

Our favorite part of the Lake Superior Zoo in Duluth is the Nocturnal Building. Upon entering, some well-lit primate habitat makes the transition to "the nocturnal loop" where the night-loving creatures live. You'll follow a horseshoe shaped passageway through a series of habitats in glass enclosures, mostly Southern Hemisphere animals, illuminated with dim, reddish light. Birds and mammals such as burrowing and spectacled owls, fruit bats, flying squirrels, prehensile-tailed porcupine, two-toed sloth, swift fox, Pallas cat, kinkajou, and large-spotted genet, complete the roster. Of these, only the Pallas cat (named for the first person to describe the species, Peter Simon Pallas,

The Nocturnal Building is alive with Southern Hemisphere night life. (Photo: Greg Brick)

based on his trip through Siberia from 1768 to 1774) is listed as actually denning in caves. But the darkness conveys a cave-like atmosphere generally.

One question regards the very nocturnality of the exhibit. The animals are kept in darkness during the day, when they are most active, so their "day" is made to be during our night. Otherwise they could not maintain their natural diurnal cycle. Another distinction could be made between crepuscular and nocturnal animals; the former are most active at twilight.

The zoo has many other attractions of course. Low-lying parts of the Lake Superior Zoo were devastated by the 2012 flood, when Kingsbury Creek overtopped its banks, but the upper buildings survived unscathed. The Griggs Learning Center in the main building, along with café and gift shop, a barnyard and petting zoo, butterfly house, big cat habitat, and the Australia/Oceania Building, are connected by looping pathways.

Directions:	7210 Fremont Street, Duluth, MN.
Seasons/Hours:	Year-round, 10 a.m. to 5 p.m. Fee.
Length:	100 feet.
Precautions:	N/A
Amenities:	Café and gift shop, a barnyard and petting zoo.
Information:	Lake Superior Zoo, 7210 Fremont Street, Duluth, MN 55807; 218-730-4500, www.lszooduluth.org.

HIGHWAY 61 TUNNELS
LAKE COUNTY

Its rocky cliffs, pebble beaches, and bold headlands have drawn visitors to the North Shore for years. Public and private development—including thousands of acres of parkland, miles of trails, and comfortable lodging—have added to the list of attractions.

This rugged, spectacular shoreline was shaped by millions of years of geologic action. More than one billion years ago, lava flows rose from a magma chamber that was located where Lake Superior is today. You can see evidence of these flows in the Sawtooth Mountains that approach the Lake in northern Lake and Cook Counties and in the basalt rock along the shoreline—for example, at Palisade Head, a mile south of Tettegouche State Park, and at Shovel Point, located in the Park just northeast of the mouth of the Baptism River. Palisade Head offers particularly awe-inspiring views, and on a clear day you can see the westernmost of Wisconsin's cave-riddled Apostle Islands, about 25 miles away.

Long after the lava had hardened, glaciers moving over the area, scoured and carved the outcrops, cliffs, and gorges. Today, streams and rivers in the region move swiftly toward Lake Superior, often whirling through rapids and plunging over waterfalls.

No wonder visitors come to view the power and beauty of this craggy landscape. A weekend spent hiking, fishing, or just plain exploring here can vanquish the most serious case of modern-age stress. Deadlines? Forget them. Visitors quickly replace any sense of being overwhelmed at the workplace with a sense of overwhelming wonder at the majesty of the North Shore.

All visitors begin their North Shore adventures from U.S. 61, which runs north from Duluth all the way to Thunder Bay, Ontario. Along the way, it traverses towns built for commerce. One hundred years ago, towns like Two Harbors, Tofte, and Grand Marais primarily served the mining, lumber, and fishing industries. Today, of course, visitors do their part to keep Minnesota's economy healthy.

The approximately 150-mile stretch of 61 from Two Harbors north to Grand Portage is so stunning that it is one of less than 20 other routes designated an All-American Road by the Federal Highway Administration. The designation marks highways so scenic that they are destinations in themselves. The route is also one of Minnesota's Scenic Byways.

Driving on this route north from Two Harbors, visitors soon come to the Silver Creek Tunnel, one of two engineering feats designed to make the highway safer for the growing number of travelers that tour here annually. Before the tunnels, Highway 61 contained several sharp and steep curves that sometimes made for hazardous driving. To make matters worse, rock slides along the eroding shoreline occasionally closed the road.

Completed in 1995, Silver Creek Tunnel extends 1,400 feet through the solid cliff. About five miles north, the 854-foot Lafayette Bluff Tunnel also improves safety for North Shore visitors. Both tunnels are convex, curving with the bluff and away from the lake, and they enhance rather than detract from the beauty of the rocky terrain. In fact, the Lafayette Bluff Tunnel received the Seven Wonders of Engineering Award from the Minnesota Society of Professional Engineers.

Thanks to the tunnels, Highway 61 now gives safe access to the majestic North Shore. In addition to lodges, marinas, and beaches, the road leads travelers to several of the state's most popular parks—Tettegouche, Gooseberry Falls, and Split Rock Lighthouse, to name a few.

Directions:	Reach U.S. Highway 61 from I-35 in Duluth.
Seasons/Hours:	Year-round.
Length:	The Silver Creek Tunnel is about 1,400 feet long. About five miles north, the Lafayette Bluff Tunnel is more than 800 feet in length.
Precautions:	N/A
Amenities:	Where to begin? The North Shore is home to magnificent state parks, homey lodges, and recreational opportunities for visitors of all tastes—including antique collectors, fishermen, mountain bikers, cross-country ski enthusiasts, and rock hunters.
Information:	Explore Minnesota, Northeast Region 800-438-5884. Website: www.exploreminnesota.com; search for North Shore Scenic Drive. Also see: www.onlyinyourstate.com/minnesota/unique-tunnel-mn/.

CORUNDUM QUARRY, SPLIT ROCK LIGHTHOUSE STATE PARK
LAKE COUNTY

Anorthosite is sometimes loosely called "moon rock" by geology students because the Apollo lunar landings brought back samples that proved to be the same mineral as found here on Earth. The famous Split Rock Lighthouse, which has served as a beacon to generations of sailors on Lake Superior, sits atop a cliff made of this mineral. And it early drew the attention of what would become the 3M Corporation, the M's standing for Minnesota Mining and Manufacturing.

The early 3M prospectors mistook the anorthosite for the valuable abrasive corundum, used in making sandpaper and so forth. In 1902, they dug several pits and left hardware that can be seen along the Gitchi-Gami Trail in this state park. Watch for the signage as you hike south from the Split Rock lighthouse toward Corundum Point, which has great views of the lake.

According to the Minnesota Department of Natural Resources, cliffs here are "masses of anorthosite, an unusual igneous rock … made of the light-colored mineral plagioclase." About 1.1 billion years ago, large blocks of anorthosite were carried up from deep below the surface in molten, dark-colored diabase. At the same time dark basalt lava flows formed much of the Park's bedrock. Beginning about two million years ago, according to the DNR, "a series of glaciers scraped across the landscape, scouring out the Lake Superior basin and molding the hills and valleys of the uplands. Finally, water filled the now ice-free basin, and streams eroded the dramatic river valleys."

The lighthouse has an interesting history in itself. Completed in 1909, the lighthouse followed a not-uncommon Lake Superior tragedy. Beginning in 1899, the Merrill and Ring Lumber Company logged Norway and white pine from the area and operated a short railroad up the river. Pilings from the old wharf and dam are still visible jutting out of the water at the mouth of the river. In 1905, a Lake Superior November gale claimed one barge and six ships within a dozen miles of the Split Rock River. The tragedy sparked the construction of the lighthouse. For 59 years, the lighthouse keepers warned ships away from the rock and treacherous North Shore with its 370,000-candlepower beacon. The federal government deeded the lighthouse to the state in 1971 and five years later, the Minnesota Historical Society assumed operation of one of the most photographed lighthouses in the nation. Its unique Fresnel lens is a rarity among lighthouses.

Directions:	Split Rock State Park is on U.S. Highway 61 northeast of Twin Harbors.
Seasons/Hours:	Year-round, 8 a.m. to 10 p.m. Park fees.
Length:	About 1.5 miles along the Gitchi-Gami Trail.
Precautions:	The unpaved trail is of moderate difficulty in hilly terrain.
Amenities:	Split Rock Lighthouse State Park offers campgrounds.
Information:	Split Rock Lighthouse State Park, 3755 Split Rock Lighthouse Road, Two Harbor, MN 55616. Phone: 218-595-7625. Website: www.dnr.state.mn.us/state_parks/index.html and use the Park Finder.

NORTH SHORE CAVES. TETTEGOUCHE STATE PARK
LAKE COUNTY

The relentless pounding of the waves of Lake Superior against Minnesota's North Shore has gnawed away at rock joints that are suitably oriented, carving out caves, especially in Lake County. Sometimes they are called "sea caves," but on a freshwater lake that expression seems rather odd, so they are referred to as littoral (shoreline) caves by geologists. Strangely, no relic littoral caves from the much higher post-glacial shorelines of Lake Superior have been found. All the known caves formed at present-day lake levels.

Aside from swimming in Lake Superior's frigid waters (not recommended) there are two ways you can see the North Shore's caves: you can visit them by cruise boat or kayak

CAVE OF WAVES NEAR ILLGEN CITY ON LAKE SUPERIOR.

Cave of Waves on the shore of Lake Superior is beloved by kayakers. (Postcard from the Greg Brick Collection)

or you can spot them from certain observation points on the shore. Unlike Wisconsin's Squaw Bay caves, which you can hike or ski into during severe winters that freeze the bay, the North Shores caves are inaccessible to hikers in any season. The ice here never freezes solid enough to permit a two-legged approach.

Still, the adventurous can safely visit the caves by kayak. If you don't own a kayak, area outfitters can help. The Lake Superior Water Trail, partially complete, will eventually circumnavigate the lake. Already, campsites and rest areas are readily available to North Shore kayakers.

Less daring explorers can visit by cruise boat (try North Shore Scenic Cruises near Beaver Bay) or view the caves from several vantage points on land. Look for observation points near bays and inlets or where the shoreline zigzags sharply like a broken pane of glass. These observation stations provide the angle you need to peer into the caves.

Certainly the best known cave along the North Shore is the **Cave of the Waves**, featured on postcards. It formed in a rocky promontory of rhyolite forming one side of Crystal Bay in Tettegouche State Park, near the town of Silver Bay, Minnesota. Follow the Shovel Point Trail down to the rocky cove. The cave is most often visited by kayakers but can also be reached by wading out from shore if you have a wetsuit, as Greg was able to do one fine day when the lake surface was glassy smooth and calm. Standing inside the watery cave, 100 feet long and elbow-shaped, he had his own private vista of Lake Superior.

The **Thunder Caves** at the mouth of the Manitou River are nearly as well known, also being depicted on postcards. Once a tourist attraction, the caves are now private. Other easily accessed shoreline caves include the **Iona Beach Cave**, near a beautiful beach of pink shingle at the Iona's Beach Scientific and Natural Area (SNA), and the **Two Harbors Cave**, in a park in the city of that name.

Directions:	Tettegouche State Park is on U.S. Highway 61, 4.5 miles northeast of Silver Bay.
Seasons/Hours:	Year-round, 8 a.m. to 10 p.m. Park fees.
Length:	The Shovel Point Trail at Tettegouche State Park is 2 miles in length. The Park features more than 20 miles of hiking trails and in winter maintains 12 miles of cross-country ski trails and another 12 miles of snowmobile trails.
Precautions:	The Shovel Point Trail is of moderate difficulty.
Amenities:	Tettegouche State Park offers campgrounds and cabins for year-round use. In the summer you can rent canoes to explore the inland waterways, and in winter snowshoes are available.
Information:	1) Tettegouche State Park, 5702 Highway 61, Silver Bay, MN 55614. Phone: 218-226-6365. Website: www.dnr.state.mn.us/state_parks/index.html and use the Park Finder. 2) North Shore Scenic Cruises on Silver Bay Marina, near Beaver Bay; website: https://northshoresceniccruises.com/

PAULSON MINE
COOK COUNTY

Test pits of the short-lived Paulson Mine lie off the Gunflint Trail roughly forty-five miles from Grand Marais. The underground iron mine opened in 1888 on the Gunflint Range, separated from the Mesabi Range by an intrusion of the Duluth Gabbro. To transport the ore, the Port Arthur, Duluth and Western Railway, first chartered in 1883 as the Thunder Bay Colonization Railway, was extended to Ely and the mine in order to link with the Duluth and Iron Range Railroad. The new line opened in January 1893, just before the mine closed, due to the growing competition of Mesabi Range mines and the Panic of 1893.

The group Wild North Minnesota has erected a marker on the Gunflint Trail commemorating the Paulson Mine. The sign reads, in part:

> Iron ore was discovered in the Gunflint area in 1886 by pioneer prospector Henry Mayhew. Mining began when the American Realty Company, headed by Minneapolis banker, John Paulson, opened the underground Paulson Mine in 1888. Without rail, or even passable roads, horse-drawn sleighs were used to haul mine equipment and supplies over frozen winter terrain.

Located within the Superior National Forest, remnants of the Paulson Mine were revealed by the Ham Lake Fire in 2007. Two years later, a new hiking trail, the Centennial Trail, was opened to mark the 100th anniversary of the Forest Service. The 3.3-mile loop trail goes right past several test pits.

Also of interest in the vicinity: Chik-Wauk Museum & Nature Center contains various mining exhibits, along with information on logging and other early area industries.

Directions:	Access the Centennial Trail from the parking area at the Kekekabic Trailhead approximately 46 miles from Grand Marais on the Gunflint Trail (Cook County Road 12). To reach Chik-Wauk Museum & Nature Center from Grand Marais, follow the Gunflint Trail (Cook County Road 12) 55 miles, to Moose Pond Drive (Cook County Road 81). Turn right and drive a quarter mile to the entry gate.
Seasons/Hours:	Year-round.
Length:	N/A
Precautions:	Challenging hiking trail.
Amenities:	N/A
Information:	National Geographic, Heart of the Continent website: www.traveltheheart.org and search for Paulson Mine.

DEVIL'S KETTLE, JUDGE MAGNEY STATE PARK
COOK COUNTY

Along the Brule River, where it flows through Judge Magney State Park, is the famous Devil's Kettle. Half the river vanishes as a waterfall into a 50-foot-deep pothole eroded into rhyolite bedrock. One guess was that there must be a secret lava tube more than a mile long that conveys this water to Lake Superior. While long dubbed a mystery, even by the prestigious Smithsonian Institution, the vanishing water is really only an optical illusion, as was proved by careful stream gauging by the DNR above and below the pothole in 2016. The water does not travel by a secret passage, but rather rejoins the Brule River in the immediate vicinity. Non-recovery of objects thrown into the pothole proves little, as they get caught up in eddies and woody debris under the waterline. The irony is that stream gauging has also revealed that many North Shore rivers unexpectedly do decrease in volume as they flow downstream, a phenomenon known as "Surber's Paradox," after fisheries biologist Thaddeus Surber (1871-1949), who first described it in 1922. The water sinks into rock crevices but where it remerges is uncertain. The hunt for the missing springs is ongoing.

Directions:	Judge Magney State Park on U.S. 61 about 15 miles north of Grand Marais. MN.
Seasons/Hours:	Year-round, 8 a.m. to 10 p.m., but approaching the lip of the kettle at night or in winter could be dangerous. Park fees.
Length:	N/A
Precautions:	Steep, uneven trail uphill for a distance of one mile. No onsite ranger in winter.
Amenities:	Campground, picnic tables.
Information:	Judge Magney State Park, 4051 East U.S. 61 Grand Marais, MN 55604. Phone: 218-387-6000. Website: www.dnr.state.mn.us/state_parks/index.html and use the Park Finder.

The mystery of the vanishing water at Devil's Kettle has been solved.
(© State of Minnesota, Department of Natural Resources, reprinted with permission)

ISLE ROYALE NATIONAL PARK
HOUGHTON COUNTY, MICHIGAN

Technically, Isle Royale National Park is in Michigan, though it lies close to Minnesota, less than 25 miles east of Grand Portage in Lake Superior. You can reach the western edge of the island in less than two hours by cruise boat. Isle Royale is also accessible by private boat or seaplane. No motor vehicles, bicycles, or pets are permitted on Isle Royale or the smaller islands surrounding it. For overnight stays you can camp or stay at the Rock Harbor Lodge.

Isle Royale contains a small cave, the remains of several 19th-century copper mines, and pre-Columbian mining pits. For thousands of years before European explorers arrived on the scene, early Native Americans mined copper using hammerstones to dislodge chunks and hammer them into tools.

The bedrock of the archipelago is basalt, formed more than one billion years ago, when a huge rift appeared in the earth's crust and multiple lava flows left behind a series of ridges. Signs of more recent glaciation remain in deposits in the southwest section of the Isle.

A 40-mile trail runs along the most prominent ridge, Greenstone Ridge, which traverses almost the length of the island. Various trails lead to Wendigo Mines, Island Mine, Isle Royale Mine, Minong Mine, Siskowit Mine, Suzy's Cave, and several inland lakes and overlooks. Available on the park website, its annual newspaper, *The Greenstone*, features a map showing trails and points of interest.

Directions:	Isle Royale is accessible by private boat or seaplane or by the boats of the Grand Portage Isle Royale Transportation Line: https://www.isleroyaleboats.com/.
Seasons/Hours:	Open from mid-April through October 31. Full transportation services available from mid-June through Labor Day.
Length:	The park archipelago is 45 miles long. The hike to Suzy's Cave from Rock Harbor is 3.8 miles.
Precautions:	Many hiking trails are quite challenging. The temperature is cool: the average daily high in July is 68; the average daily low in July is 50. As you hike and camp about the island, water must be carried or boiled. Be prepared for mosquitoes, biting flies, and yellow jackets. Bring medications and a first aid kit. No cars, bikes, or pets allowed.
Amenities:	Hiking, kayaking, boating, scuba diving.
Information:	Isle Royale National Park, 800 East Lakeshore Drive Houghton, MI 49931. Phone: 906-482-0984. Website: www.nps.gov/isro/.

THE QUEST FOR LAVA TUBES

Both Hawaii and Iceland are slathered with a generous coating of basaltic lava flows honeycombed by lava tubes. These natural tubes form while the lava is still hot, serving to drain away the excess molten rock in a flaming river that sometimes reaches the sea with a loud, steaming hiss. Once the lava has cooled and solidified, the tubes are safe for humans to explore, containing features that mimic limestone caves, such as "lavacicles" (lava stalactites). And since Minnesota's North Shore has the same kind of lavas, wouldn't you expect to find lava tubes there, too, even though they cooled more than a billion years ago?

Several generations of Minnesota cave explorers (including one of the authors of this book!) have certainly thought so and set off to find them. However, to this day no Minnesota lava tubes have been identified.

However, you can get your lava tube cravings satisfied at the Minnesota Zoo in Apple Valley, MN! Along the Northern Trail you'll find "Russia's Grizzly Coast" complete with live bears. As you hike the trail you pass through an artificial lava tube. More at: https://mnzoo.org/blog/animals/russias-grizzly-coast/

Russia's Grizzly Coast is the best evocation of a lava tube in Minnesota. (Photo: Minnesota Zoo)

FURTHER READING

Iowa Underground, Greg Brick, 2004

Minnesota Caves: History & Lore, Greg Brick, 2017

Minnesota's Geology, Richard Ojakangas and Charles Matsch, 1982

Minnesota's Iron Country: Rich Ore, Rich Lives, Marvin G. Lamppa, 2004

Minnesota Underfoot, Constance Jefferson Sansome, 1983

Roadside Geology of Minnesota, Richard Ojakangas, 2009

Seven Iron Men: The Merritts and the Discovery of the Mesabi Range, Paul de Kruif, 2007

Subterranean Twin Cities, Greg Brick, 2009

Wisconsin Underground, 2nd edition, Doris Green, 2019

ABOUT THE AUTHORS

Doris Green follows her curiosity about topics such as education, genealogy, and the natural environment, writing for local, regional, and national publications. This diversity sometimes leads to insights not found in more focused approaches.

This second edition of *Minnesota Underground* resulted from a conversation with the publisher of her fourth nonfiction book for general audiences, *Elsie's Story: Chasing a Family Mystery*, which traced its origin to a childhood trip to Wisconsin's Cave of the Mounds.

Green launched and co-published *Wisconsin Community Banker* magazine for the former Community Bankers of Wisconsin and was a communications specialist with the School of Human Ecology at the University of Wisconsin-Madison. She also served as a publisher at Magna Publications, which then published books, as well as newsletters for college and university audiences.

She holds a bachelor's degree from the University of Wisconsin-Madison School of Education and a master's degree from the School of Journalism and Mass Communication. She lives with her husband, Michael H. Knight, and three distracting cats in a log home near Spring Green, Wisconsin. Reach her at dorisgreenbooks.com.

Greg Brick Ph.D. was employed as a hydrogeologist at several environmental consulting firms and has taught geology at local colleges and universities. He has edited the *Journal of Spelean History* since 2003. He has published more than 200 articles about caves and was the recipient of the 2005 Cave History Award from the National Speleological Society. His first book, *Iowa Underground: A Guide to the State's Subterranean Treasures*, was published by Trails Media Group in 2004. His second book, *Subterranean Twin Cities*, published by the University of Minnesota Press in 2009, won an award from the American Institute of Architects. His third book, *Minnesota Caves: History and Lore*, was published by The History Press in 2017 and was nominated for the Minnesota Book Awards. His work has been featured in *National Geographic Adventure* magazine as well as on the *History Channel*. He has led guided tours of caves for the Minnesota Historical Society and the University of Minnesota College of Continuing Education. He established the Minnesota Spring Inventory at the DNR and mapped thousands of springs around the state. At present he's editing a multi-authored volume on the Caves and Karst of the Upper Midwest. Reach him at gregbrick.org.

www.ingramcontent.com/pod-product-compliance
Lightning Source LLC
Chambersburg PA
CBHW042116190326
41519CB00030B/7519